MatWerk

Edited by
Dr.-Ing. Frank O. R. Fischer (Deutsche Gesellschaft für Materialkunde e.V.)
Frankfurt am Main, Deutschland

Die inhaltliche Zielsetzung der Reihe ist es, das Fachgebiet „Materialwissenschaft und Werkstofftechnik" (kurz MatWerk) durch hervorragende Forschungsergebnisse bestmöglich abzubilden. Dabei versteht sich die Materialwissenschaft und Werkstofftechnik als Schlüsseldisziplin, die eine Vielzahl von Lösungen für gesellschaftlich relevante Herausforderungen bereitstellt, namentlich in den großen Zukunftsfeldern Energie, Klima- und Umweltschutz, Ressourcenschonung, Mobilität, Gesundheit, Sicherheit oder Kommunikation. Die aus der Materialwissenschaft gewonnenen Erkenntnisse ermöglichen die Herstellung technischer Werkstoffe mit neuen oder verbesserten Eigenschaften. Die Eigenschaften eines Bauteils sind von der Werkstoffauswahl, von der konstruktiven Gestaltung des Bauteils, dem Herstellungsprozess und den betrieblichen Beanspruchungen im Einsatz abhängig. Dies schließt den gesamten Lebenszyklus von Bauteilen bis zum Recycling oder zur stofflichen Weiterverwertung ein. Auch die Entwicklung völlig neuer Herstellungsverfahren zählt dazu. Ohne diese stetigen Forschungsergebnisse wäre ein kontinuierlicher Fortschritt zum Beispiel im Maschinenbau, im Automobilbau, in der Luftfahrtindustrie, in der chemischen Industrie, in Medizintechnik, in der Energietechnik, im Umweltschutz usw. nicht denkbar. Daher werden in der Reihe nur ausgewählte Dissertationen, Habilitationen und Sammelbände veröffentlicht. Ein Beirat aus namhaften Wissenschaftlern und Praktikern steht für die geprüfte Qualität der Ergebnisse. Die Reihe steht sowohl Nachwuchswissenschaftlern als auch etablierten Ingenieurwissenschaftlern offen.

It is the substantive aim of this academic series to optimally illustrate the scientific fields "material sciences and engineering" (MatWerk for short) by presenting outstanding research results. Material sciences and engineering consider themselves as key disciplines that provide a wide range of solutions for the challenges currently posed for society, particularly in such cutting-edge fields as energy, climate and environmental protection, sustainable use of resources, mobility, health, safety, or communication. The findings gained from material sciences enable the production of technical materials with new or enhanced properties. The properties of a structural component depend on the selected technical material, the constructive design of the component, the production process, and the operational load during use. This comprises the complete life cycle of structural components up to their recycling or re-use of the materials. It also includes the development of completely new production methods. It will only be possible to ensure a continuous progress, for example in engineering, automotive industry, aviation industry, chemical industry, medical engineering, energy technology, environment protection etc., by constantly gaining such research results. Therefore, only selected dissertations, habilitations, and collected works are published in this series. An advisory board consisting of renowned scientists and practitioners stands for the certified quality of the results. The series is open to early-stage researchers as well as to established engineering scientists.

Herausgeber/Editor:
Dr.-Ing. Frank O. R. Fischer (Deutsche Gesellschaft für Materialkunde e.V.)
Frankfurt am Main, Deutschland

Denise Reichel

Temperature Measurement during Millisecond Annealing

Ripple Pyrometry for Flash Lamp Annealers

With a Preface by Dr.-Ing. Wolfgang Skorupa

 Springer

Dr. Denise Reichel
Dresden, Germany

Dissertation, TU Bergakademie Freiberg, 2015

MatWerk
ISBN 978-3-658-11387-2   ISBN 978-3-658-11388-9 (eBook)
DOI 10.1007/978-3-658-11388-9

Library of Congress Control Number: 2015957376

Springer
© Springer Fachmedien Wiesbaden 2015

Printed on acid-free paper

Springer is a brand of Springer Fachmedien Wiesbaden
Springer Fachmedien Wiesbaden is part of Springer Science+Business Media
(www.springer.com)

"Nothing has such power to broaden the mind
as the ability to investigate systematically and truly
all that comes under thy observation in life."

Marcus Aurelius (121-180)

Preface

This book is the result of a dissertation work at the Helmholtz-Zentrum Dresden-Rossendorf based on a contract with the centrotherm AG at Blaubeuren/Germany. The background is the need to find out more sophisticated approaches for measuring temperatures in the time range below one second down to the millisecond range. The need for that results from the most advanced Silicon technology and other applications using modern lamp-based annealing techniques for thermal treatments of substrates where the activation of electrical, optical and other functions is needed at negligible impurity diffusion and/or thermal load to the substrate.

Denise Reichel guides the reader from an excellent review, also published in a renowned review journal, to sophisticated experiments based on ripple pyrometry. The final concept was proved for the case of flash lamp annealing of shallow Boron-dopants in Silicon. This well-selected example was over many years one of the key technology drivers of advanced silicon technology. Moreover, this annealing technology gets more and more attention from the side of low-cost large area electronics where functional layers on temperature-sensitive substrate material have to be treated. The development of temperature measurement on the millisecond scale as described here will have an important impact on the further development of advanced thermal processing of modern materials.

Helmholtz-Zentrum Dresden-Rossendorf   Dr.-Ing. Wolfgang Skorupa

Acknowledgement

I wish to express my sincere gratitude to Dr. W. Skorupa, Dr. W. Lerch and J. C. Gelpey for the idea of this topic and their supervision throughout the work. I cannot express in a few words the contribution, which their experience and willingness to advise has made to my professional standing and development. Special thanks should be given to Prof. Dr. Dirk C. Meyer for taking the supervision of my doctoral thesis at the Technische Universität Bergakademie Freiberg. Also, I am greatly indebted to Dr. M. Voelskow, Dr. L. Rebohle, Dr. H. Stöcker, Dr. V. Heera, Dr. K.-D. Bolze, Dr. H. Rick, Dr. R. A. Yankov, Dr. P. Pichler, Dr. M. Hackenberg, Mr. T. Schumann and Mrs. R. Rietzschel for fruitful discussions and valuable assistance in the preparation of this PhD Thesis.

Moreover, I would like to acknowledge the support by the Department of Technical Research at the Helmholtz-Zentrum Dresden-Rossendorf.

Besonderer Dank gilt meinen Eltern und allen Menschen, die mich seit jeher unterstützt haben.

Dresden   Denise Reichel

Abstract

Millisecond thermal processing is a powerful tool for reducing dopant diffusion after implantation in a semiconductor material while the electrical activation is comparable to conventional annealing time scales. Further, it is needed to protect temperature sensitive substrates used in up-to-date thin foil applications. For detailed studies as well as closed-loop process control, one of the major challenges is to find suitable temperature measuring approaches and methods.

Temperature measurement during millisecond annealing is particularly difficult regarding the high light energy density flash pulses used, which create considerable background radiation. This thesis is devoted to a new technique for temperature measurement of semiconductor wafer surfaces in millisecond annealers. It uses the idea of ripple pyrometry, which is known from conventional rapid thermal processing, where the amplitude modulation of the circuit-driven halogen lamps due to the net frequency is used for background correction. To the best of the author's knowledge there is no alternative to background-corrected temperature measurement during millisecond annealing without the use of filters, which require extensive cooling.

The here presented method provides for the first time ripple pyrometry to millisecond annealing by using an opto-mechanical approach via voice coil technology. It will be demonstrated that amplitude modulation has been achieved for light pulses with a duration of 20 ms and that the modulation frequency can be multiplied by a suitable choice of the system parameters to modulate light pulses with a duration as short as 1 ms. Further, this thesis proves that the amplitude modulation is reproducible for equal light intensity and that it can be simulated knowing the shape of the incident light pulse, which makes this method also controllable.

Temperature measurement has been achieved with an uncertainty > 15 K at a temperature of 1300 K and > 23 K at a temperature of 1650 K by online reflectivity measurement through comparing the modulation depth of the incident light pulse before and after reflection off the wafer surface. These results are up to eight times more accurate in the determination of the temperature than offline alternatives. Moreover, methods of sensitivity analysis were applied to identify the influence of various physical and system parameters. Own raytracing simulations have been performed to study the determination of the sample reflectivity by ripple pyrometry. It has been shown that the detector position, the angle and the chamber geometry play an important role. In particular, it could be found that the detector position is crucial for reliable reflectivity measurements and it has been found to meet the real sample reflectivity best in close proximity of the wafer surface.

One of the major advantages in contrast to other methods is that ripple pyrometry works independently of the user having any knowledge about the material. The information on the material – state and phase – is already contained in the online reflectivity measurement, which makes this technique free of calibration and particularly suitable for solid-liquid phase changes during annealing.

Table of Contents

List of Figures

List of Tables

List of Abbreviations, Acronyms and Special Terms

A_i	dopant in interstitial state
A_s	dopant in substitutional state
acc. to	according to
AI	dopant interstitialcy
a.u.	arbitrary units
amplitude modulation	periodic change of the amplitude to mark the carrier wave
annealing	heat treatment in order to heal implantation damage and for electrical activation of the dopants
as-measured reflectivity	the reflectivity that is determined by ripple pyrometry regardless of the true value
AV	dopant-vacancy pair
BED	Boron enhanced diffusion
BIC	Boron-interstitial-cluster
Bragg condition	condition for the reflection of waves off the crystal lattice
C	capacity
ε	emissivity
emissivity	material-specific ability for radiative transfer relative to a black body
FA	furnace annealing
FLA	flash lamp annealing
FWHM	full width half maximum
h	Planck constant
heating lamps	lamp bank to serve the purpose of heating the wafer to be annealed
I	interstitial defect
I_{bb}	black body radiation intensity in a.u.

intrinsic ripple	amplitude modulation of the light output following the alignment of coils and capacitors in the LC circuit
ion implantation	irradiation of solids by particle beam for doping
IV	Frenkel pair
k_B	Boltzmann constant
L	inductivity
L_D	diffusion length
LA	laser annealing
λ	wavelength
lamp detector	detector which faces the lamp bank
LC circuit	resonant circuit consisting of coils with inductivity L and capacitors with capacity C
M_{bb}	black body energy density in W/m^2m
MSA	millisecond annealing
N_A	Avogadro constant
offline	measurement outside the FLA process
online	measurement during the FLA process
oscillation frequency	set frequency of the voice coil movement
pyrometer	radiation detector for elevated temperatures
R	reflectivity
R_L	radiance
radiation thermometry	temperature measurement using radiation detectors
range straggling	statistical uncertainty of particle range following implantation
ripple	here: amplitude modulation on lamp radiation
ripple frequency	amplitude modulation frequency on flash pulse
ripple pyrometry	temperature measurement by measuring the sample reflectivity through amplitude modulation of the radiation of the heating lamps
ripple size	modulation depth of the amplitude modulation
RT	room temperature

RTA	rapid thermal annealing
RTP	rapid thermal processing
SIMS	secondary ion mass spectrometry
SiB_n	Silicon boride
SiO_2	Silicon dioxide
SiO_4	Silicon tetraoxide
solid phase regrowth	recrystallization of an amorphised solid surface
SPR	solid phase regrowth
SRP	spreading resistance profiling measurement of resistance as a function of depth from which the carrier concentration profile is deduced
T	temperature
τ	transmissivity
t_A	flash pulse duration = annealing time
t_R	thermal response time
TED	transient enhanced diffusion
thermal mass	physical inertia of a material upon an increase of temperature
thermal profile	temperature evolution over time
ultra-shallow implantation	implantation at low energies (single-digit to sub-1 keV regime)
UV	ultraviolet (regime of the electromagnetic spectrum)
V	vacancy
ν	frequency
VIS	visible (regime of the electromagnetic spectrum)
wafer detector	detector which faces the wafer
w.r.t.	with respect to

1 Introduction and Motivation

Microelectronic industries are mainly based on the semiconductor technology – a fact that can be explained by the possibility to vary the electric current through a semiconductor by a control circuit. This in turn is based on the ability to dope a semiconductor material positively or negatively. Doping involves ion implantation of suitable elements from a higher (n-doping) or a lower (p-doping) main group of the periodic table and subsequent thermal annealing of the material. The annealing serves two purposes. For one, it minimises the implantation-induced damage of the crystal structure. Second, the dopant atoms need to be electrically activated by increasing the internal energy to move them from interstitial sites into substitutional lattice sites.

There are different methods for thermal annealing which can be distinguished by the annealing time into four main categories: furnace annealing (FA, minutes up to hours), rapid thermal annealing or processing (RTA or RTP, seconds), millisecond annealing (MSA) and laser annealing (LA, nanosecond time scale) where MSA and RTA are lamp-based methods.

Besides the desired processes which lead to electrical activation of the dopants and defect annealing after implantation, thermal annealing also causes undesired dopant diffusion. For further downscaling of microelectronic devices and with that for shallow junction design both lateral and transient dopant diffusion processes need to be suppressed. For annealing of non-patterned wafers, however, transient diffusion is of particular interest and will be called simply *diffusion* in the text that follows. This is a suitable approximation as experiments for the present work have been carried out on uniformly doped substrates. As it is a function of time and temperature, one possibility to limit diffusion is by reducing the annealing time. Having in mind the Arrhenius equation, however, this requires a rise in temperature for an exponential increase in the process rate to obtain comparable electrical activation of the dopants over a shorter timescale (Atkins & de Paula, 2013). Although undesired diffusion is also enhanced at elevated temperatures, short annealing cycles are still beneficial if they are small with respect to the thermal conduction within the material. Ultra-short-time annealing in the order of milliseconds and nanoseconds leads to transverse temperature gradients across the wafer which hinders diffusion into the bulk material and leaves the backside of the wafer at an intermediate temperature. For FLA – a type of MSA, this can and will be of great interest with regard to many applications that are based on temperature sensitive substrates as well as with regard to thermal annealing of shallow junctions. For scanning techniques such as laser annealing, the localised thermal gradient may introduce stress into the material and leads to wafer breakage. Although wafer breakage is also known for FLA, the stress introduced into the material is less significant (Sedgwick, 1983).

The competing processes of dopant activation and diffusion as functions of time and temperature require a reliable control of the process temperature for semiconductor technology. However, temperature control during short-time annealing requires a contactless means of measurement. Conventionally, radiation thermometry is the method of choice for online temperature measurement and process control. In the literature (Reichel, 2011) an extensive review thereto can be found. However, as indicated above, short-time annealing (RTA, MSA, LA) is based on optical energy input by halogen lamps, gas discharge/ flash lamps or lasers in the order of decreasing annealing time. If spectral overlap between the detection regime and the annealing spectrum cannot be avoided, which is often the case for RTA and MSA, this causes considerable background radiation that needs to be separated from the signal of thermal radiation.

In RTA a range of approaches to temperature measurement exist which, however, are bound to specific local conditions, e.g. Raman Spectroscopy, or which cannot be readily adapted to shorter timescales though time resolution is nowadays no longer an issue (Reichel, 2011). One promising approach is the use of intelligent filter designs (Stuart, 2002), (Walk & Theiler, 1994). However, for increasing incident energy density, absorption of light by the filter material may not be negligible anymore and re-emission needs to be considered.

Another idea for the solution to this problem arises from a method used in RTA for emissivity-corrected temperature measurements: *ripple pyrometry* [6]. In RTA it is based on electrical modulation of the background radiation from the halogen lamps which is used for online reflectivity measurement of the material to be annealed. The reflectivity obtained is a measure of the contribution of the background radiation towards the overall detection signal. From this information a background-corrected temperature reading is gained. Online reflectivity measurement can further account for phase changes during the annealing process. However, in contrast to halogen lamps, the optical output of flash lamps during MSA cannot be electrically modulated. This is because flash lamps are driven by a capacitor discharge. Thus, a new technique has been looked for, which can be adapted to ripple pyrometry for millisecond annealing.

In the year 1990, a paper has been published based on the "First Workshop on Subsecond Thermophysics" which has been held in Gaithersburg, Maryland, USA two years earlier (Cezairlizan, 1990). This paper illustrates the need for millisecond temperature control.

Researchers from well-known institutes in the USA and Italy discussed "Issues and Future Directions in Subsecond Thermophysics Research". Their discussions referred to thermophysical processes for timescales between 1 s and 10^{-12} s and the determination of thermophysical properties away from thermal equilibrium. The authors note that especially the study of the liquid phase will be of interest in future. In fact, the phase change that is related to melting can be detected using ripple pyrometry and thus, the temperature reading can be corrected accordingly.

They also recognize the issues associated with process control through radiation thermometry which is due to the strong background radiation during short-time annealing. They stress that the measurements outside thermal equilibrium are very important on the one hand, but also difficult to perform and investigate regarding temperature gradients, difficulties in determining properties at high temperatures as well as changing optical properties. A method has been looked for that can account online for phase changes and reliable temperature measurement irrespective of the material or complex calibration procedures during subsecond annealing.

Due to the strong background light of the flash lamps, conventional radiometry / pyrometry is not suitable for temperature determination. Ripple pyrometry provides a solution to eliminate the influence of the reflected flash onto the pyrometer signal and simultaneously it provides online information on phase changes of the material. To the best of the author's knowledge ripple pyrometry could not have been applied for millisecond annealing, yet.

According to current industrial demands the present thesis intends to investigate in depth the suitability of online reflectivity determination for true temperature measurement during millisecond annealing. Furthermore the technical challenges and possibilities for ripple pyrometry in FLA devices will be studied in the following thesis.

Chapter 1 introduces the issue of temperature measurement during millisecond annealing. Chapter 2 describes the technical and physical aspects of a flash lamp annealer as the representative of millisecond annealing. It also considers defect creation, migration as well as defect annealing and compares flash lamp annealing (FLA) to other means of thermal processing.

Chapter 3 considers the fundamentals of thermal radiation and surface temperature measurement and highlights the limitations of radiation thermometry for FLA.

Chapter 4 reviews prior methods for surface temperature measurement and discusses the current state of the art. Chapter 5 provides the reader with a detailed insight into the background and idea for the present work. It demonstrates the application of ripple pyrometry for FLA purposes and explains the new technique which has been developed in the course of the present work. Information on its concept and set-up as well as on alternative approaches is provided and its suitability for temperature measurements in the case of millisecond flash pulses is studied. An overview on the methods used for the present work is given.

Chapter 6 concentrates on the applicability of the technique presented for surface temperature measurement during FLA for low and elevated temperatures. For this purpose, a series of experiments on shallow Boron-doped and undoped Silicon wafers have been carried out and the temperature measurement uncertainty has been investigated.

Chapter 7 closes this thesis with conclusions pertinent to the temperature measurements carried out during millisecond annealing using the presented new concept.

2 Fundamentals of Flash Lamp Annealing of Shallow Boron-Doped Silicon

MSA of semiconductors is usually performed using flash lamps. It has been shown that FLA holds the balance between effective dopant activation and a low diffusion rate into the bulk [8], [9]. Moreover, it allows for thermal processing of surfaces and interfaces independently of the substrate material. Furthermore, in contrast to laser annealing, FLA enables one to perform thermal treatment of large-scale samples without increasing the process time.

2.1 Electrical and Optical Characteristics of Flash Lamps

Figure 1 shows the general set-up of an FLA chamber. A set of halogen lamps preheats the sample back to prevent thermal stresses and shock in the wafer, which may cause breakage due to the temperature gradients occurring as a consequence of the millisecond heating.

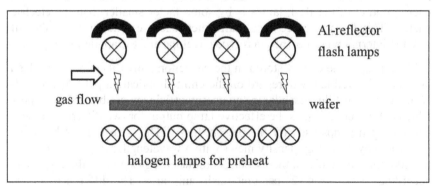

Figure 1:   Schematic drawing of the set-up of an FLA apparatus.

The halogen and flash lamps, which are filled typically filled with Xenon or Argon, are protected by a 3 mm thick quartz plate from possible evaporation of gases from the sample. A gas flow – usually of Nitrogen or Argon – ensures an inert atmosphere. Above the flash lamps Aluminium reflectors are mounted to decrease the loss of light intensity for annealing. The round shape of the reflectors is chosen such that the flash light is collimated onto the wafer surface. The flash lamps can be operated between pulse times of

130 µs and 80 ms at full width half maximum (FWHM). The maximum energy density on the surface amounts to approximately 150 J/cm² which is limited by the capacitance of the lamp circuit. The housing of the chamber is made of Aluminium which melts at a temperature T of 933.47 K. However, due to the high reflectivity of Aluminium the flash cannot heat it up to its melting temperature. Still water cooling is supplied to the chamber walls and to the Aluminium reflector during the preheating step.

An LC resonator consisting of coils with inductance L and capacitors with capacitance C creates flash pulses with a time duration $t = \frac{1}{f}$, whereby

$$f = \frac{1}{2\pi\sqrt{LC}} .\tag{1}$$

The electric energy

$$E = \frac{1}{2}CU^2\tag{2}$$

that is stored in the capacitors can be calculated knowing the capacitance C and the charging voltage U, whereas the discharge of the capacitors is retarded by the presence of the coils as a function of inductance. The optical energy output of the flash lamps is by some factor smaller than the electric energy. This factor is a technical parameter that is characterised by the fill gas, the internal pressure, the glass type envelope and the applied voltage.

With respect to the energy stored in the capacitors, an efficiency of the flash lamps can be defined with regard on the emitted optical output. Theoretically, using Xenon-filled flash lamps an energy efficiency of about 50 % is possible [10]. Figure 2 shows the effective lamp output for two different devices and the flash lamps used for the present thesis. The measurement has been performed by an energy density meter with a wavelength range from 400 nm to 1400 nm. The depicted theory considers the energy stored in the capacitors according to Equation (2) per unit flash lamp area. The differences in the absolute values of energy density for the two chamber devices represent different geometries / lamp bank areas. These results reproduce the theoretical upper limit of 50 %. For flash pulses with time durations of 20 ms and 3 ms at FWHM at a low charging voltage, a significant deviation from this theoretical value cannot be found.

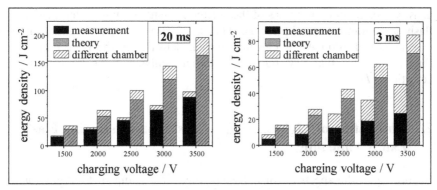

Figure 2: Comparison of the theoretical (predicted by Equation (2)) and measured energy densities for flash pulse times of 20 ms (left) and 3 ms (right) at FWHM.

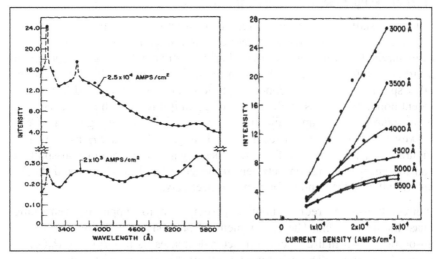

Figure 3: Increase in the UV output of a Xenon-filled flash lamp as a function of wavelength and current density. Left: The Xenon spectrum is shown for two different current densities. Right: Peak position on the wavelength scale and intensity for growing current density (figure taken from [11]).

Due to recombination and collision of free electrons with bound positive Xenon or Argon ions in the high pressure tube, the emission spectrum of the flash lamps is continuous as shown in Figure 3. High current density increases the optical output in the UV/Vis region whereas for low current density spectral emission lines of the excited Xenon/Argon ions dominate the flash lamp output (cf. Figure 3). The steep rise of peak intensity at a wavelength of 3000 Å for elevated current densities in the right graph of Figure 3 shows that UV output is strongly enhanced for large current densities [11].

2.2 Absorption of Light in Single-Crystal Silicon during Flash Lamp Annealing

According to the spectrum of a Xenon-filled flash lamp in Figure 3 for a material to be suitable for millisecond annealing two prerequisites need to be fulfilled. First, it needs to be absorbing in the visual part of the electro-magnetic spectrum and second, the time scale of annealing should be in the order of the light thermal response time [2].

Silicon crystallises in a face-centred cubic structure with a two atom base at $\{(0,0,0)\ (\frac{1}{4}, \frac{1}{4}, \frac{1}{4})\}$ (diamond structure). Each Silicon atom is the heart of an even tetrahedron with four sp^3 hybrid orbitals stretching out to the corners of a virtual tetrahedron. In the direction of each hybridised orbital there is an-other sp^3 orbital from a neighbouring Silicon atom to overlap and form a co-valent bond. The corresponding band diagram in Figure 4 shows that Silicon is an indirect band gap semiconductor with a bandgap energy at 1.12 eV. However, Silicon also has local direct bandgaps $E_{\Gamma 1}$ and $E_{\Gamma 2}$ as shown in Figure 4. E_{SO}, E_X and E_L refer to the energy gap due to spin-orbital-splitting and the energy separation between the valence band maximum and the X- and L-valley of the conduction band, respectively.

Indirect transitions require phonon participation to supply the remaining quasi-momentum $\hbar k$. The requirement of phonon participation for optical absorption in an indirect semiconductor decreases its absorptivity $\alpha_0(\lambda)$ as shown in Figure 5. The absorption coefficient at the indirect bandgap E_g is comparatively low with respect to its value at the direct bandgaps $E_{\Gamma 1}$ and $E_{\Gamma 2}$ for all temperatures. As a guidance to the eye the absorption spectrum in the vicinity of the indirect band gap is plotted on a logarithmic scale (see inset in Figure 5).

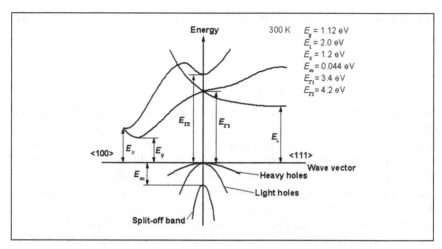

Figure 4:   Band structure of single-crystal Silicon (figure taken from [12]). Eg = indirect bandgap; EL = energy gap between the valence band maximum and the L-valley; EX = energy gap between the valence band maximum and the X-valley; ESO = energy gap due to spin-orbital splitting; EΓ1, EΓ2 = local direct bandgaps

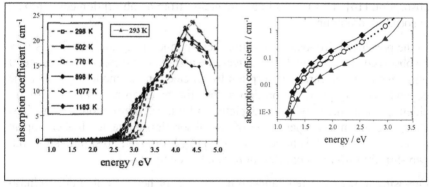

Figure 5:   Absorption spectrum for single-crystal Silicon as a function of photon energy and temperature (acc. to [13]). The weak rise in absorptivity around the global indirect bandgap E_g at 1.12 eV at a temperature of 293 K is depicted in the detail on the right hand side (data taken from [14]) along with $\alpha(\lambda, T)$ for two different temperatures according to Equation (3) for comparison (legend of the overall absorption spectrum applies). To prove comparability, the absorptivity $\alpha_0(\lambda)$ at a temperature of 293 K according to Green and Keevers [14] is also shown in the graph on the left hand side in the overall absorption spectrum.

For temperatures T between 298 K and 1183 K and photon energies below $E_{\Gamma1}$ Sun et al. found that absorptivity α can be described by

$$\alpha(\lambda, T) = \alpha_0(\lambda)\exp\left(\frac{T}{T_0}\right), \qquad (3)$$

whereby $T_0 = 703$ K [13]. The inset in Figure 5 also shows that the drop in absorptivity around the indirect bandgap at a photon energy of 1.12 eV is shifted for elevated temperatures towards decreasing photon energy, i.e. increasing wavelength, according to the bandgap-narrowing with rising temperature. In addition to optical absorption across the bandgap, free carrier absorption or intraband absorption has an impact on the absorption spectrum. The slope in absorptivity at high photon energies, i.e. at short wavelengths, is weaker for elevated temperatures, which may be attributed to a higher reflectivity due to free carrier absorption above the direct bandgap [13]. As will be shown later (cf. subsection 2.3.3), shallow doping leads to large dopant concentrations despite medium implantation doses [15]. Therefore, heavy doping needs to be considered in the following. Figure 6 shows the absorptivity for lightly and heavily doped Silicon [16]. Figure 6 and Figure 7 exhibit the absorption spectrum for increasing dopant (Boron) concentration. The slope of the graphs can be well explained by enhanced free carrier absorption [16]. This results in a conversion from an absorption edge into an absorption minimum.

In the presence of (dopant) impurities zero-phonon interband transitions may be observed. To investigate this effect, the authors subtracted the influence of intraband and free carrier absorption by an electron gas model to deduce the dependence on optical absorption across the bandgap as a function of incident photon energy which is depicted in Figure 7. Zero-phonon transitions should result in a steeper slope of the absorption curve for higher dopant concentrations which is merely the case [16]. Moreover, no significant bandgap shift for different concentrations can be found.

Opposing to the expected bandgap narrowing for heavy doping due to intraband levels, there is a concurrent process known as the Burstein-Moss effect as a consequence of the shift of the Fermi energy level into the valence band for heavy p-type doping. Partial band-filling of the valence band with holes leads to a higher photon energy necessary to raise a bound electron from the valence band into the conduction band although this effect is more pronounced for n-type doping. These opposite effects lead to a so-called band gap renormalisation [16], [17].

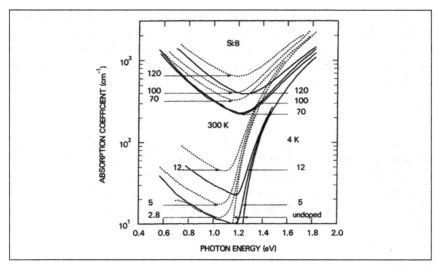

Figure 6:   Absorption spectrum for various dopant concentrations (in units of 10^{18} cm^{-3}) measured at temperatures of 4 K and 300 K (figure taken from [16]).

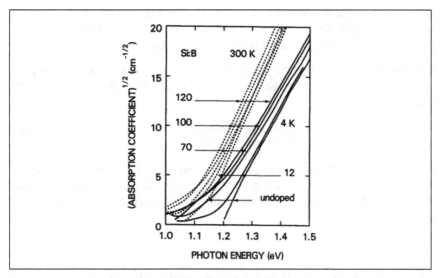

Figure 7:   Square root of the absorption coefficient to show optical absorption across the bandgap. Free carrier absorption has been removed by a model based on interacting electron-gas from Figure 6. Concentrations are given in units of 10^{18} cm^{-3} (figure taken from [16]).

Incident photons are primarily absorbed through electron-hole-formation via the bandgap. Excessive photon energy is subsequently released to the lattice by electron-phonon-coupling. The resulting change in thermal energy dQ during a temperature interval dT at constant volume V or constant pressure p is defined as heat capacity

$$C_{V,p} = \left|\frac{dQ}{dT}\right|_{V,p}.$$ (4)

There are two results for C_V depending on the model that is applied: For the calculation of the mean thermal energy Q Einstein viewed the atoms of a semiconductor lattice as harmonic oscillators moving independently with the same frequency ν_E at temperature T. Debye, however, considered phonons in a box with a frequency spread $d\nu$. The general description of Q can be expressed by

$$Q = \langle n(E)\rangle E, \text{ where}$$

$$\langle n(E)\rangle = \frac{1}{\exp\left(\frac{h\nu}{k_B T}\right)-1}$$ (5)

refers to the mean occupation number according to the Bose-Einstein statistics, E indicates the phonon energy, k_B the Boltzmann constant, h indicated Planck's constant and ν represents the frequency.

Following Einstein's theory, where the lattice atoms are treated as independently moving harmonic oscillators with equal frequency ν_E, the heat capacity

$$C_V = 3\,N_A h\nu_E \frac{1}{(exp\left(\frac{h\nu_E}{k_B T}\right)-1)^2} exp\left(\frac{h\nu_E}{k_B T}\right)\frac{h\nu_E}{k_B T^2}.$$ (6)

is obtained using Equation (5) to perform Equation (4), whereby N_A represents the Avogadro constant. Debye's approach considers phonons as a gas in a box with a volume V and with a frequency range $d\nu$ to give a thermal energy of

$$Q = 3\frac{V^3 \nu^2}{2\pi^2 c_S^3}\int_0^{\nu_D} E\,d\nu$$ (7)

whereby ν_D denotes the Debye frequency and c_S the phonon propagation speed in the solid.

Thus, for Debye's theory C_V is expressed by

$$C_V = 9Nk_B \left(\frac{T}{\Theta_D}\right)^3 \int_0^{\Theta_D} \frac{\frac{h\nu}{k_B T}^4 \exp\left(\frac{h\nu}{k_B T}\right)}{\left(\exp\left(\frac{h\nu}{k_B T}\right) - 1\right)^2} \tag{8}$$

In the high temperature limit both results approach $3N_A k_B$ (= 24.9 Jmol^{-1}K^{-1}) at a temperature which is known as the Debye temperature $\Theta_D = \frac{\hbar\omega_D}{k_B}$. This upper limit agrees with the value due to the Dulong-Petit law which is derived from the equipartition theorem according to which each mode in a monoatomic solid carries the energy $Q = 3N_A k_B T$. Equation (4) thus yields a heat capacity of $3N_A k_B$ as found above. For Silicon this limit corresponds to a heat capacity of 0.89 Jg^{-1}K^{-1} at a temperature of 636 K [18].

Knowing C_V, the heat capacity at constant pressure C_p can be obtained from the thermodynamics for cubic solids. For a solid having a linear expansions coefficient α, an isothermal compressibility κ_T and a volume V, C_V and C_p differ at temperature T by

$$C_p - C_V = \alpha^2 V \, T \frac{1}{\kappa_T}. \tag{9}$$

The difference in absolute values is negligible. However, in contrast to C_V, C_p continues to increase above the Debye temperature, which is mainly due to the fact, that the coefficient of linear expansion for Silicon as the compressibility of solids can be neglected.

Experimental results for C_V and C_p are shown in Figure 8. The dotted line indicates estimates due to the Dulong-Petit law. At low temperatures C_p and C_V are almost indistinguishable. At elevated temperatures, however, C_p violates this law and increases above the Debye temperature.

The distribution of temperature T in space x and time t is described by the differential heat equation

$$\frac{\partial}{\partial t} T(x,t) - \frac{\lambda}{\rho c} \frac{\partial^2}{\partial x^2} T(x,t) = 0 \tag{10}$$

for the one-dimensional case, whereby λ_c specifies the thermal conductivity, ρ the material density and c the specific heat capacity of the material. The specific heat capacity of Silicon has already been looked at. Figure 9 shows experimental data for the thermal conductivity of single-crystal Silicon at temperatures between 2 K and 2000 K [20]. The contributions to the thermal conductivity of Silicon can be distinguished into photon (= radiative contri-

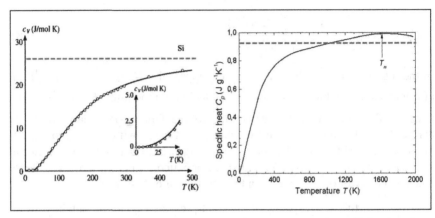

Figure 8:   Experimental data for the heat capacity at constant volume (left, taken from [18]) and at constant pressure (right, taken from [19]) for single-crystal Silicon in comparison to the result predicted by the Dulong-Petit law (dotted line).

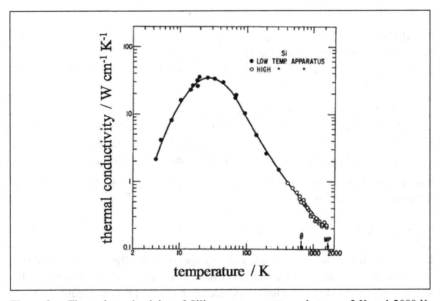

Figure 9:   Thermal conductivity of Silicon at temperatures between 2 K and 2000 K. The results have been produced with two apparatus for the low (solid circles) and high temperature regime (open circles), (figure taken from [20]).

bution, < 2 %), phonon and electronic participation [20]. The latter can further refer to electron-hole-pairs (bipolar) or single carriers (polar) taking part in the temperature propagation. The decrease for elevated temperatures can be explained by increasing phonon-phonon-scattering.

Solving Equation (10) delivers a thermal gradient which is formed across the wafer thickness for ultra-short annealing cycles. The amount of heat

$$dQ = \frac{\lambda_c}{d} A (T_t - T_b) dt \qquad (11)$$

transferred between two parallel surfaces of a wafer with thickness d, surface area A and a thermal conductivity λ_c at temperatures T_t and T_b for the bottom and top surface, respectively, during an annealing time dt, can be described by the Fourier's law in Equation (11). For short flash pulses the time provided for heat diffusion into the bulk, defined by the thermal conductivity λ, is limited and therefore a larger thermal gradient can form compared to RTP. As a desired consequence, this gradient supports heat transport into the bulk through conduction after the flash pulse time is over. Thus, the bulk serves as a heat sink for rapid cooling of the surface and therefore limits undesirable dopant diffusion from the surface into it (cf. Equation (18b)).

The thermal diffusivity

$$a = \frac{\lambda}{\rho c} \qquad (12)$$

for the temperature range between room temperature (RT) to 1400 K is shown in Figure 10. In order that a thermal gradient can be formed, as discussed above, the thermal response time of the wafer

$$t_R = \frac{L^2}{a} \qquad (13)$$

should be in the order of – or larger than – the annealing flash pulse duration t_A, whereby L can be understood as the wafer thickness and a as the thermal diffusivity. Comparing Equation (13) and Figure 10, a thermal gradient can form in a 520 μm (~ L) thick wafer for single-digit annealing times in the order of 1 ms and for the entire temperature regime.

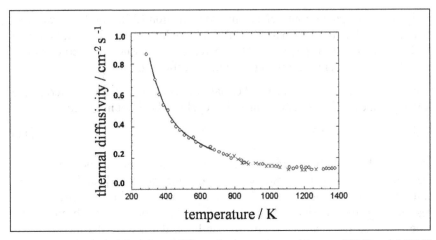

Figure 10: Thermal diffusivity of Silicon for temperatures between 293 K and 1400 K (figure taken from [21]).

2.3 Defect Annealing and Dopant Activation in Boron-Doped Silicon

2.3.1   Types of Defects in Single-Crystal Silicon

Defects in Silicon can be divided into three main groups: native defects, implantation-related and impurity-related defects. Native defects, in turn, are classified into four categories: zero-dimensional or point defects (interstitialcies, interstitials and vacancies), one-dimensional or line defects (edge and screw dislocations), two-dimensional or area defects (grain, tilt and twin boundaries as well as stacking faults) and three-dimensional or volume defects (voids and precipitates) [22]. As point defects play an important role during annealing, they will be closer looked at in the text that follows.

Vacancy-type defects are empty lattice sites. Interstitials, in contrast, refer to non-substitutional atoms, i.e. they do not occupy a regular lattice site; and interstitialcies denote a pair of non-substitutional atoms occupying a substitutional lattice site as a group. There are also extended interstitialcies which can be understood as a non-matching number of Silicon atoms and lattice sites [23]. The most important extended defect for dopant diffusion is the rod like {311} defect which originates from Silicon-interstitial clustering.

This defect plays an important role in Boron diffusion in Silicon and will be considered in more detail later in this chapter.

A native point defect can be either isolated or bonded to a dopant atom. The interaction of these two species is also important for the diffusion and activation mechanisms during annealing. Following thermodynamical considerations and regardless of external energy absorption, native point defects form during thermal equilibrium in a real Silicon lattice as opposed to an ideal lattice. There are different processes responsible that are described by Fahey in [23].

A Silicon atom may spontaneously leave its lattice site becoming an interstitial-type defect (I-type) leaving behind a vacancy (V-type) to form a Frenkel pair (IV-type):

$$I + V \leftrightarrow IV$$

Then, one of the two may migrate to combine with the opposite surface defect such that a regular lattice site is reoccupied by a Silicon atom. By this process energy is emitted which can be uptaken by another lattice atom to create again a vacancy and an interstitial. The formation of defects is further supported during crystal growth at the liquid-solid interface. Huitema and van der Eerden describe a "competition of generation of defects at the growing surface and 'partial' healing in a wider interface region" [24]. They found by Monte Carlo simulations that a healing depth d exists above which the growth process itself cannot heal the defects that are produced at the liquid-solid-interface. In this case, extended defects are formed for which a minimum defect size can be defined that relates linearly to the depth where the defect is generated. The type of defects formed have been shown to depend upon the relation between the growth rate V and the temperature gradient at the liquid-solid-interface ΔT. At a large $V/\Delta T$ ratio, vacancies are the dominant defect species whereas for a low $V/\Delta T$ ratio, interstitials occur in excess. Agglomerates of point defects arising from supersaturation upon cooling of the melt are known as microdefects or swirl defects [25].

A further source for point defects is the variety of impurities. Despite of diffusion, impurities may consume or create point defects or form more complex structures like the Johnson complex, an impurity-vacancy-complex. While dopants are desired impurities, the most important representatives of undesired impurities in terms of concentration are Oxygen and Carbon atoms in the called order which are mainly introduced during crystal growth by the silica crucible and the graphite heater, respectively. Oxygen is a crucial impurity at concentrations between 10^{17} cm^{-3} and 10^{18} cm^{-3}. It is assumed to be incorporated into the Silicon primarily as an interstitial bound to two neighbouring Silicon atoms forming neutral complexes. At a temperature of 723 K, however, SiO$_4$ complexes, so-called thermal donors, are formed at a maximum concentration of 10^{16} cm^{-3}. Due to supersaturation of Oxygen inside Silicon upon cooling of the melt, dielectric SiO$_2$ precipitates form [26], [27].

The incorporation of Carbon into the lattice is mostly substitutional at a concentration of about 10^{16} cm^{-3}. Presumably due to the lattice mismatch caused by substitutional Carbon and the resulting strain release, when forced into an interstice, Carbon is an efficient trap for Silicon interstitials. There is a mutual interaction forming Carbon-Oxygen complexes due to Oxygen being larger and Carbon being smaller than Silicon, thereby reducing internal lattice strain. Moreover, experimental results indicate that thermal Oxygen donors, which normally form at a temperature of 723 K, also appear at elevated values in the presence of Carbon [26], [27].

As mentioned above, dopants are desired impurities. Doping of semiconductors is necessary to increase their electrical conductivity. The introduction of dopants into the material can be undergone during crystal growth or by a process called ion implantation, which itself introduces defects in the crystal lattice. When a highly energetic ion enters the material, lattice atoms are displaced from their sites. Replacement collisions are less likely. Interstitial Silicon atoms play an important role for diffusion during thermal annealing [23]. Defects produced formed by a non-amorphising ion implantation process can be either of vacancy- (V), substitutional- (A_s), interstitial- (I, A_i) or interstitialcy-type (AI) character. These defects may interact manifold [23]:

$$A_s + V \leftrightarrow AV$$

$$A_s + I \leftrightarrow AI, A_i$$

$$A_i + V \leftrightarrow A_s ,$$

where AI refers to a dopant interstitialcy of substitutional and interstitial defects, AV denotes a dopant-vacancy pair and Ai describes a lattice atom occupying an interstitial lattice site.

The mean projected range $\bar{R}_p$ during implantation of N ions travelling a distance x strongly depends on the ion energy E. It can be described by a linear differential equation following the directional diffusion model of Biersack [28]

$$\bar{R}_p = \frac{1}{N}\Sigma_i x_i \tag{14a}$$

$$\frac{d\bar{R}_p}{dE} = \frac{(2E - \mu S_n \bar{R}_p)}{(2S_t E)}, \tag{14b}$$

where S_n and S_t express the nuclear and total stopping power, respectively, and μ denotes the atomic mass ratio $\frac{M_i}{M_m}$ of the implanted and matrix atoms of masses M_i and M_m, respectively. For shallow junction engineering implanttation energies in the single-digit or sub-1 keV regime have become a subject of interest. It has been shown that for this ultra-low energy range new effects with respect to defect annealing have to be considered. Due to the decrease in the mean projected range and reduced range straggling for low-energy implantation, the defective region in the material is thinner and is positioned closer to the surface. The proximity to the surface aids in defect removal. However, due to the shallow defective region, the concentration of dopants and defects increases quickly with increasing implantation doses [15].

Boron implantation into single-crystal Silicon creates point defects, which accumulate to {311}-type extended defects. For an energy of 500 eV Boron doping in single-crystal Silicon additional Boron-interstitial clusters (BICs) have been observed [15], [29]. Their influence on the thermodynamics during annealing will be considered in the section that follows.

2.3.2   Point Defects in Silicon and Defect Migration during Thermal Treatment

Following thermodynamics at any temperature above zero Kelvin, defects occur to minimize the free energy of the real lattice. The equilibrium concentration c for a point defect species X

$$\frac{c_X}{c_{Si}} = DOF * \exp\left(\left(\frac{S_{disorder}}{k_B}\right)\right) \exp\left(\frac{-H_f}{k_B T}\right) \tag{15}$$

where c_X denotes the point defect concentration, c_{Si} the concentration of lattice sites, DOF describes the species' degree of freedom, $S_{disorder}$ is the entropy of disorder and H_f the enthalpy of defect formation [23]. This equation shows that the equilibrium concentration of X increases with growing temperature as defect creation is an endothermic process. It has been mentioned above that in the presence of dopants there may be an attractive interaction among point defects and dopants forming complexes. Thus, the total point defect concentration c_X^{tot} under extrinsic doping conditions is the sum of the equilibrium concentration c_X and the concentration of those point defects bound in complexes c_{AX}.

Annealing of point defects in Silicon is mediated by their diffusion towards the surface. Near the surface, defect dissolution and subsequent annihilation is enhanced [30]. Dopant activation is also a diffusion supported process. As mentioned before, after implantation, dopants often occupy interstitial sites. At these positions they may function as electron traps and thus decrease free carrier conduction [31]. To become electrically active, they need to be re-located onto lattice sites. At elevated temperatures vacancies are generated, which facilitate the movement of the interstitial dopant atoms onto lattice sites. This leads to electrical activation of the dopants. The thereby created intra-bandgap donor/acceptor levels shift the Fermi level towards the conduction (n-doping) / valence (p-doping) band [32].

Dopant diffusion with diffusivity D due to a concentration gradient $\frac{\partial c}{\partial x}$ are described by Fick's first (16) and second law (17) [32]

$$J = -D \frac{\partial c}{\partial x} \tag{16}$$

$$\frac{\partial c}{\partial t} = -\frac{\partial J}{\partial x} \tag{17}$$

accounting for the initiation of the diffusive flux J and its evolution in time, respectively. Combining random walk – describing a stochastical process with independent, stationary growth – and Fick's law the diffusion coefficient D is obtained. D can be expressed in terms of the diffusion length L_D and the diffusion time t_D by the Einstein-Smoluchowski relation [32]

$$D = \frac{L_D^2}{2 t_D}. \tag{18a}$$

The dependence of the diffusion coefficient

$$D = D_0 \exp\left(\left(\frac{-H_m}{k_B T}\right)\right) \tag{18b}$$

on the temperature T follows an exponential behaviour, whereby D_0 represents a temperature-independent preexponential and H_m the migration enthalpy [**32**]. There are three major diffusion mechanisms by which a dopant atom can migrate in a Silicon matrix [**23**]:

The vacancy diffusion mechanism describes the migration of a dopant atom A on a substitutional site by exchanging positions with a neighbouring vacancy V. However, this requires at least partial dissociation of the AV pair in contrast to a pure exchange of positions, which would not lead to diffusion. In fact, for the diamond-type of structure of the Silicon crystal lattice, this requires a line of travel of at least three lattice sites between each jump (cf. Figure 11).

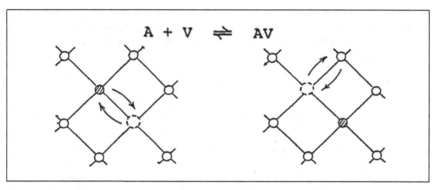

Figure 11: Illustration of the vacancy diffusion mechanism in Silicon. The vacancy needs to travel at least three lattice sites between each swap of places with the dopant for the latter to move forward (figure taken from [**23**]).

If an interstitial I is bound to a substitutional dopant atom A to yield a dopant interstitialcy AI, it may dissociate and recombine with the dopant to form a new interstitialcy AI by which the dopant may diffuse along the lattice. This mechanism is called interstitialcy diffusion mechanism and is illustrated in Figure 12.

The third main mechanism is referred to as interstitial diffusion mechanism and may be initiated by displacement of a substitutional dopant atom A_S into an interstice by an interstitial Silicon atom I. The migration stops if A ends up on a lattice site by removal of a substitutional Silicon atom from its place into an interstitial position. This mechanism, which is also referred to as kick-out mechanism, is shown in Figure 13.

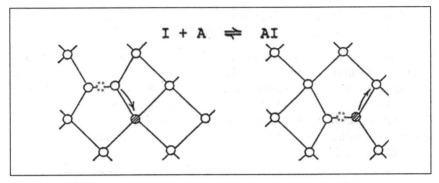

Figure 12: Illustration of the interstitialcy diffusion mechanism in Silicon. The dotted circles indicate the lattice site of the diamond structure (figure taken from [23]).

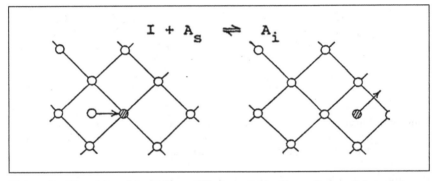

Figure 13: Illustration of the kick-out diffusion process. The exchange of places between a self-interstitial I and a substitutional dopant atom A_s initiates the interstitial diffusion through the crystal (figure taken from [23]).

A diffusion coefficient can be defined describing the diffusion of a species as a linear combination of the coefficients of each diffusion mechanism w.r.t. the respective concentrations c.

For the vacancy (cf. Figure 11) and interstitialcy mechanism (cf. Figure 12) this would result in the diffusion coefficient of species X [33]

$$D_X = D_X^V \left(\frac{c_X^V}{c_X}\right) + D_X^I \left(\frac{c_X^I}{c_X}\right). \tag{18c}$$

From this the fractional contribution

$$f_V = \left(\frac{D_X^V}{D_X}\right)\left(\frac{c_X^V}{c_X}\right) \text{ and}$$

$$f_I = \left(\frac{D_X^I}{D_X}\right)\left(\frac{c_X^I}{c_X}\right) \tag{18d}$$

of each mechanism towards the overall diffusion process of the species X can be deduced. For Boron, f_I has been shown to be unity, which means that the diffusion rate for Boron is directly proportional to excess Boron interstitials [33].

2.3.3   Enhanced Diffusion in Shallow Boron-Doped Silicon

It is commonly agreed that the diffusion process in Silicon for both, n-type and p-type implants, takes place via an enhanced mechanism, also known as transient enhanced diffusion (TED) which is faster for higher annealing temperatures [34]. Literature sources attribute TED in Boron-doped Silicon to interstitial diffusion (cf. Figure 13), and investigations into the subject have shown that extended defects are formed by excess Silicon interstitials after ion implantation which may precipitate into rod-like, also called {311} defects according to the respective crystal plane, to decrease the free energy of the system [33], [35]. The origin for TED lies in the re-emission of Silicon interstitials from the extended {311} defect. Through {311} defect generation the local equilibrium between interstitial and substitutional dopants is shifted and the thus enhanced dopant diffusivity is in direct relation to the interstitial supersaturation [23], [34], [35]. The diffusion takes place via the interstitial mechanism (kick-out diffusion) explained above [36]. This leads to dramatic consequences w.r.t. the junction depth. Even for shallow implantation, Boron may diffuse into the bulk to depths in the order of several hundred nanometres [33]. However, diffusion into the bulk is inhibited at elevated temperatures due to enhanced diffusion of interstitials towards the surface and their annihilation near the surface. The driving force for this process is the re-establishment of the equilibrium [33], [36], [37].

For ultra-shallow junction engineering, sub-1 keV implantation energies are required. For this regime the defective layer is in close proximity to the surface. Near the surface defect dissolution and subsequent annihilation is enhanced. Therefore, for ultra-low implantation energies less diffusion would be expected. In fact, Agarwal et al. worked out that Boron diffusion is enhanced for the sub-1 keV regime by a factor of 3 ... 4 despite the proximity of the defective layer to the surface. The authors call this phenomenon Boron-enhanced diffusion (BED). In fact, the contribution of BED is stronger than the anomalous temperature behaviour of TED [15], [29]. Already at medium doses ultra-shallow Boron implantation results in considerably high Boron concentrations within the shallow defect layer. The concentration of Boron in the defective layer is assumed to reach an atomic concentration of around 6 at.% at its peak for an implantation energy of 500 eV and at an implantation dose of 10^{15} cm^{-2} (cf. Figure 71). This results in BICs which may further precipitate into a Silicon boride (SiB$_n$) phase. It is assumed that the SiB$_n$ injects interstitials into the bulk [15], [29], [38]. The diffusivity enhancement by the formation of BICs hinders the formation of {311} defects and providing a more efficient source for interstitial point defects as {311} extended defects [29]. Despite the low energy implant, BED increases the junction depth undesirably. However, at low annealing temperatures an uphill diffusion of Boron towards the surface has been observed, which raises the interest in Boron implants for shallow junctions again [39]. Moreover, Boron has been shown to be a better choice for ultra-shallow junction formation than Arsenic, despite its lower solubility and higher diffusivity [15], [40].

2.4 Comparison of Flash Lamp Annealing with Furnace, Rapid Thermal and Laser Annealing

The driving force for short-time annealing is maintaining the same efficiency in dopant activation as conventional furnace annealing while having suppressed dopant diffusion. Short-time annealing comprises a time scale of 10^{-9} s – 1 s. It can be classified by the relation of annealing time scale t_A to thermal response time of the wafer t_R (cf. Equation (13)) [2]. In this respect, RTA, where $t_A \gg t_R$, can be described as isothermal processing. The entire wafer is heated up to a certain temperature for times of seconds up to hours. The thermal stress is kept low, whereas the thermal budget is high compared with laser or millisecond annealing. The term *thermal budget* describes the temperature that a material is exposed to over a period of time.

In contrast, pulsed laser annealing, where $t_A \ll t_R$, may be described as an adiabatic process. The surface is rapidly heated by the laser above the melting point of the sample material, and the melt depth is only a function of density distributions of the laser energy.

Annealing time scaling in the order of the thermal response of the wafer can be described as thermal flux processes, where the melt depth is not only a function of input energy density, but also of wafer backside temperature. Thermal flux processes refer to scanning laser and FLA whereas the first introduces thermal stress into the wafer material by the scanning movement. FLA, in contrast, provides treatment of the wafer over its entire surface, although wafer breakage is also known for FLA [2].

The interplay between dopant diffusion and dopant activation as a consequence of short-time annealing and conventional furnace annealing reveals that millisecond annealing holds the best balance [2], [8], [9]. Figure 14a shows secondary ion mass spectrometry (SIMS) profiles for Boron doped Silicon samples at an implantation energy of 500 eV after heat treatment by RTA and FLA at a temperature of 1473 K.

The concentration profiles for Boron upon FLA and RTA mainly differ in the tail region where appreciable diffusion into the bulk takes place for RTA. However, TED is less important for FLA due to the restriction in annealing time. Figure 14b shows spreading resistance profiling (SRP) results for the same samples. As to be seen, the junction depth can be considerably reduced using FLA [8].

Pulsed laser annealing melts a sample surface which results in box-shaped concentration profiles after annealing due to the enhanced diffusivity of the dopant inside the melt. This is advantageous for junction engineering leading to a well-defined junction depth at the liquid-solid interface and may result in enhanced electrical activation compared to conventional annealing methods. However, for ultra-low energy implantation at medium doses increased lattice strain by the incorporation of the smaller Boron atom into the lattice may even result in reduced electrical activation [41], [42]. According to the distinction above non-melt continuous wave laser annealing is comparable to FLA [2]. However, the scanning mode of lasers represents a major disadvantage in comparison to FLA.

A study by Zhu et al. on defect characterization of Boron-doped Silicon samples upon FLA showed that defects are effectively removed near the surface [43]. The process conditions are comparable to those used in the present

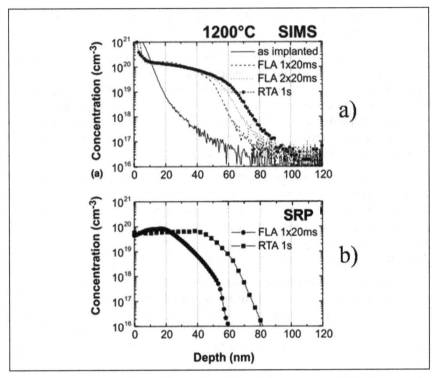

Figure 14: **a)**: SIMS profiles of Boron-doped samples at an energy of 500 eV, an atomic concentration of 1015 cm^{-2} and at a temperature of 1473 K after RTA (= RTP) and FLA. **b)**: SRP measurements to compare the electrical activation of Boron-doped Silicon samples for rapid thermal (left) and flash lamp (right) annealing (both figures are taken from [8])

work. Shallow Boron implantation at a concentration of 10^{15} cm^{-3} was performed at an energy of 500 eV at RT. FLA was subsequently applied at intermediate temperatures of 1023 K and 1373 K to a peak temperature of 1573 K. However, the annealing time was considerably shorter at 1 ms. The surface during FLA is a sink for point defects, which leads to efficient defect removal or at least to a decrease in defect cluster size upon millisecond annealing. Zhu et al. presented transmission electron microscopy measurements revealing effective defect annihilation above a temperature of 1523 K. At lower temperatures small clusters were observed in close proximity (at a depth of ca. 5 nm) to the surface [**43**].

3 Fundamentals of Surface Temperature Measurements during Flash Lamp Annealing

3.1 Planck's Law of Thermal Radiation – a Historical Outline [44]

In the year 1802, William Hyde Wollaston was the first to discover dark lines in the solar spectrum. His findings were remade twelve years later independently by Joseph Fraunhofer. Surprisingly, these lines could be reproduced in the laboratory through absorption experiments of different element flames. In a series of papers, Gustav Kirchhoff compared the emission spectra of thirty elements to the solar spectrum and formulated his famous law which relates the coefficients of thermal emission and absorption [45], [46]. Moreover, Kirchhoff's law describes their contribution towards background radiation (here: radiation at thermal equilibrium, namely black body radiation), which had been unknown at the time. In the year 1879, Josef Stefan found out that the total radiation power

$$M_{bb} = \sigma T^4,\tag{19a}$$

of a black body source, where σ denotes the Stefan-Boltzmann constant, is proportional to the fourth power of the body's absolute temperature T. Five years later, Ludwig Boltzmann provided the theoretical proof [47].

In the year 1893, Wilhelm Wien showed that M_{bb} is in fact a function of the product λT with the wavelength λ. These considerations lead him to the derivation of the spectral power density of black body radiation

$$M_{bb}^\lambda = C\,\lambda^{-5}\exp\left(\frac{-D}{\lambda T}\right)\tag{19b}$$

with the constants C and D. In the year 1911, Wien was awarded the Nobel Prize for his research into black body radiation. Equation (19b) is in agreement with the experiments. However, the error in M_{bb}^λ increases with rising λT.

Planck found by comparison of the mean emitted energy and the mean absorbed energy of a substrate that

$$M_{bb}^\nu = \frac{8\pi\nu^2}{c^3}E,\tag{19c}$$

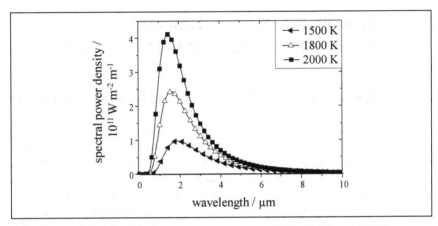

Figure 15: Black body spectrum at temperatures of 1500 K, 1800 K and 2000 K.

whereby ν denotes the frequency of the emitted radiation, c the velocity of light and E the average energy of a black body. Equation (19c) shows that the spectrum of black body radiation can be described without the knowledge of any material constants. Lord Rayleigh, an expert in the field of wave theory and author of the well-known book at that time "The Theory of Sound" [48] verified Planck's result (Equation (19c)) by considering standing waves in a box. With the knowledge of the time about statistical thermodynamics he set $E = k_B T$. The low frequency end of the black body spectrum was thus corrected.

However, the high frequency limit known as the UV-catastrophe could not be solved for, yet. A function was needed that strongly increases to a maximum and then falls steeply towards the high frequency limit (see Figure 15). Initiated by the results of two scientists from the Prussian Academy of Sciences, Planck found the full answer to the problem [49].

His law of thermal radiation with

$$M_{bb}^{\nu} = \frac{8\pi h\nu^3}{c^2} \frac{1}{\exp\left(\frac{h\nu}{k_B T}\right)-1} \text{ and} \tag{19d}$$

$$M_{bb}^{\lambda} = \frac{8\pi hc^2}{\lambda^5} \frac{1}{\exp\left(\frac{hc}{\lambda k_B T}\right)-1}, \tag{19e}$$

which will be referred to as Planck's law in the following, could now fully replicate the experimental findings of black body radiation.

3.2 Fundamentals of Thermal Radiation and Surface Temperature Measurements

Equations (19d) and (19e) describe the black body radiation spectrum in dependence on the frequency ν and the wavelength λ, respectively. Mathematical integration yields the total spectral power M_{bb} for a given temperature T. The description of M_{bb} as a function of angle ϑ and solid angle Ω is directly related to the observed radiance R_L by the Lambert cosine law

$$M_{bb} = R_L \int_0^\pi \cos\vartheta \, d\Omega. \tag{20}$$

If R_L is constant at any angle the surface is referred to as Lambertian Radiator. The ratio of R_L for a black and a real body defines the emissivity ε which is the value of utmost importance for radiation thermometry.

The excitation of phonons and plasmons with frequency $\nu_{k,p}$ and occupation number $n_{k/p}$ during the absorption process can be described as harmonic-oscillator-like behaviour. Through relaxation from the excited state into its ground state a photon energy of

$$E_{k/p} = \sum_{k/p} h\nu_{k,p} < n_{k/p}(T) > \tag{21}$$

is released. The fraction of emitted radiation in comparison to the black body spectrum is called emissivity. Emissivity can be divided into different catagories. Some of the frequently used expressions are: hemispherical (ε_h), spectral (ε_s), spectral directional ($\varepsilon_{s,d}$) and normal emissivity (ε_n). They can be mutually expressed for a constant temperature as

$$\varepsilon_h = \int \varepsilon_s(\lambda)d\lambda = \int \int \varepsilon_{s,d}(\lambda,\gamma)d\gamma d\lambda, \tag{22}$$

whereby $\varepsilon_s(\lambda, 0°) = \varepsilon_n$ and γ specifies the angle of incidence with respect to the area normal. According to Kirchhoff's law, the spectral and directional absorptivity $\alpha_{s,d}$ (λ,γ) equals $\varepsilon_{s,d}$ (λ,γ) at thermal equilibrium. For the non-equilibrium state (MSA), Kirchhoff's law is not true generally. However, it may be assumed with sufficient accuracy if the sampling time of the detector is small in comparison to the slope of the thermal profile of the sample under investigation.

A famous paper by Sato in the year 1966 shows the spectral emissivity for a temperature range from 543 K up to 1073 K (cf. Figure 16) [50]. The sample was 1770 µm thick and n-doped at a concentration of 2.94×10^{14} cm^{-3}. Sato's results have been obtained directly by comparing the radiance emitted to a black body source as well as indirectly through measurements of the as-measured reflectivity of the sample. The latter also considers partially transparent substrates due to thin samples or multiple internal reflections as opposed to the ideal situation of a complete *balance* between absorption and emission (cf. Kirchhoff's law). Figure 16 shows that emissivity drops drastically below the bandgap of Silicon, but rises again with increasing wavelength, because lattice vibration becomes dominant. Besides the shift of the bandgap towards longer wavelengths for increasing temperature, emission due to free carrier radiation sets out for elevated temperatures above 870 K raising emissivity to a stable behaviour independent of the wavelength. For highly doped Silicon the drop around the Silicon bandgap is less pronounced due to additional free carrier thermal emission [51], [52]. Other factors that influence emissivity – including surface roughness, coatings and oxide thickness – are described in detail elsewhere [53]. Measurements on directional emissivity exhibited that Silicon can be referred to as a Lambertian radiator, indeed, for angles up to 60° and only drops slightly for larger angles [54].

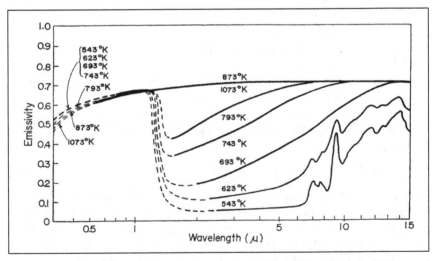

Figure 16: Spectral emissivity of single-crystal n-doped Silicon with a thickness of 1770 μm. Solid lines: direct measurement; dotted lines: indirect measurement by the determination of the reflectivity of the sample (figure taken from [50]).

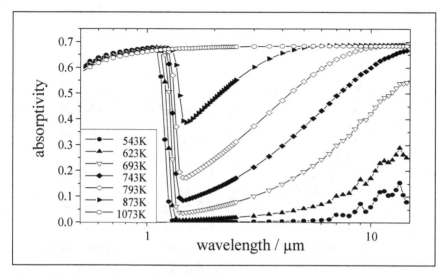

Figure 17: Absorpthion coefficient of Silicon, adapted to show the validity of Kirchhoff's law in comparison to Sato [50].

To proof the validity of Kirchhoff's law, Figure 17 is adapted as a function of wavelength to compare to Figure 16. The results have been obtained using a programme for the simulation of radiative properties with equal input parameters to the samples used by Sato for comparison [50], [55]. Clearly, Kirchhoff's law is fulfilled. In this case and if the transmissivity τ_S of the sample is negligibly small, the reflectivity $R = 1 - \alpha - \tau_S$ of its surface reduces to $1 - \varepsilon$.

Figure 18 shows the reflectivity of Silicon at wavelengths between 0.5 μm and 5 μm and for temperatures of 293 K, 693 K, 793 K and 1673 K [56]. The extremely high values below the wavelength of the bandgap of Silicon are due to free carrier absorption. Below the bandgap phonon excitation leads to increasing reflectivity for longer wavelengths and higher temperatures.

Upon melting, the reflectivity rises by a factor of two (cf. Figure 19). The data in Figure 19 has been obtained at a wavelength of 633 nm [57]. However, Figure 18 shows that the data from Figure 19 can be compared to the wavelength regime just below the wavelength of the Silicon bandgap at a

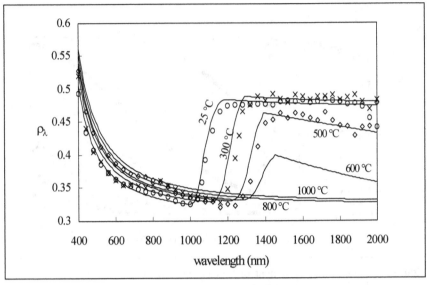

Figure 18: Spectral reflectivity of a double-sided polished Silicon wafer within the wavelength regime between 400 nm and 2000 nm. (O) experimental results at RT, (×) experimental results at a temperature of ca. 603 K, (◇) experimental results at a temperature of ca. 773 K, (—) numerical results (figure taken from [56]).

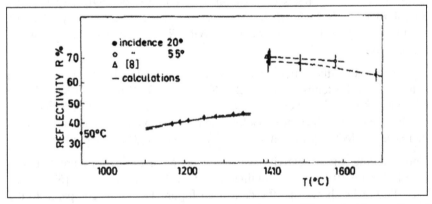

Figure 19: Reflectivity of solid and liquid Silicon for two different incident angles of 20° and 55° at a wavelength of 633 nm; comparison of literature [58] and calculated values obtained from a fourth-order polynomial fit (figure taken from [57]).

wavelength of 1100 nm as the reflectivity changes only slightly for wave-lengths from 633 nm to the bandgap. The sharp increase in reflectivity at the transition from solid to liquid Silicon, which is in good agreement with metallic behaviour [57], would result in erroneous temperature measurement if left unconsidered.

For surface temperature measurements, surely, there is a demand for the definition of a *surface*. Clearly, this depends on the penetration depth of the detection wavelength into the material (see Figure 20) as well as on the temperature depth distribution upon annealing (see Figure 21). Figure 21 shows that the latter is a strong function of annealing time as expected from Equation (11). This means, while for an annealing time of 3 ms the steep decrease in temperature within the first 100 μm of thickness into the sub-strate sets an upper limit to the surface thickness of which the temperature is measured, for flash pulse times of 20 ms at FWHM it is only determined by the penetration depth of radiation within the range of detection. Figure 20 shows the penetration depth at RT within the near-infrared wavelength regime. However, as to be seen from Figure 17 it decreases strongly as a function of rising temperature.

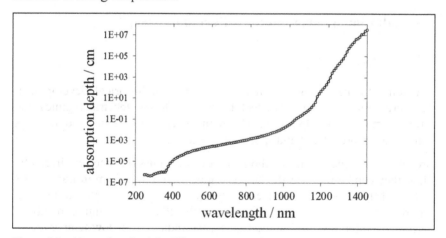

Figure 20:   Absorption depth in Silicon at RT as a function of wavelength. Data is taken from [*14*].

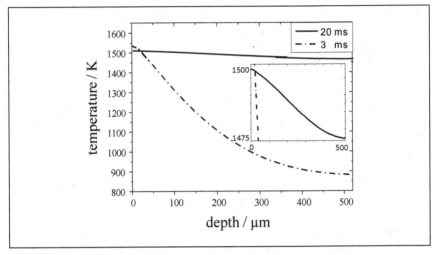

Figure 21: Depth distribution of the Silicon temperature during annealing for flash pulse durations of 3 ms and 20 ms.

3.3 Radiation Thermometry – Pyrometry [59]

3.3.1   Pyrometer Formats

According to their suggested application and the historical development, there are different types of detectors used, different spectral regimes and different methods of detection. The main discrimination of pyrometers is done between broadband and spectral pyrometers.

Broadband pyrometers detect about 90 % of the emission spectrum intensity. Thus, they can be used for the largest temperature range from just above RT to several thousand Kelvin covering the spectrum from UV/Vis to the mid-infrared regime. Due to the decrease in the Planck maximum, parabolic mirrors are often used to bundle the thermal radiation for low temperatures. As broadband pyrometers measure almost the complete thermal spectrum, the Stefan-Boltzmann law can be used to convert the readings into temperature values with sufficient accuracy.

For many applications, however, a limited detection regime is essential, which is covered by bandpass and spectral pyrometers. Presently, the most common group for reliable temperature control are semiconductor detectors, whose use is determined by the lattice and energy band structure of the semi-conductor. While broadband pyrometers can use the Stefan-Boltzmann law with sufficient accuracy for temperature measurement bandpass and spectral pyrometry require Planck's law.

More detailed information on the various detector types for broadband and spectral pyrometers as well as on other types of pyrometry can be found else-where [**59**].

3.3.2   *Limitations of Pyrometry for Flash Lamp Annealing*

The strongest contributors to limit the use of radiation pyrometry are back-ground radiation, emissivity and the shadow effect. Background radiation is especially crucial for a real-time temperature measurement. Best results are generally obtained if only one photon source is present, namely the heat source emitting thermal radiation according to its temperature. As the pyro-meter is really a photon detector it cannot distinguish between light and heat sources. It integrates over time all incoming radiation regardless of its origin. For FLA – in contrast to RTA using halogen lamps – spectral overlap in terms of a mutual range on the wavelength scale of light is negligible. For illustration, Figure 22 shows a typical flash lamp spectrum and the wave-length-dependency of commonly used annealing temperatures according to Planck's law. Intensities cannot be compared, but Table 1 indicates that the flash intensity exceeds the thermal radiation of the wafer by more than one order of magnitude despite the negligible spectral overlap. Table 1 shows the relative contribution of thermal radiation towards total irradiance of a flash pulse with a time duration of 20 ms during annealing for 30 % sample reflectivity and 70 % sample emissivity between 0.9 µm and 1.1 µm. The calculations in Table 1 do not include direct lamp radiation which ought to be avoided by suitable placing of the pyrometer. This may be accomplished by focusing the optical path of the pyrometer onto a small spectral window in the chamber wall to view the specimen.

Although it decreases for elevated temperatures, the issue of this very poor signal-to-background-ratio is serious, where signal means thermal radiation here. One solution is moving to longer wavelengths towards the peak of the Planck distributions of the processing temperatures which lie further outside the flash spectrum in the near-infrared regime between wavelengths of 1 µm and 2 µm. However, there are good reasons not to do so. Temperature can be

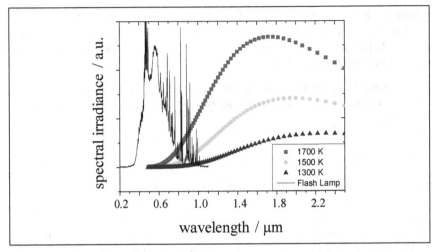

Figure 22: As-measured spectrum for Xenon flash lamps in comparison with the respective Planck curves of typical annealing temperatures (1300 K, 1500 K and 1700 K) showing the spectral overlap for the wavelength regime between 0.8 μm and 1.1 μm. Intensities cannot be compared between the optical spectrum of the flash lamps and the thermal radiation (figure taken from [59]).

Table 1: Relative contribution of thermal radiation towards total lamp irradiance.

	Temperature:			
	1300 K	1500 K	1800 K	2000 K
Relative contribution in % at a central wavelength of 1 μm ± 0.1 μm:	0.6	1.4	3.8	6.6

determined with higher accuracy at lower wavelengths of light given that, apart from the thermal radiation, there is no other photon source (at the same wavelength regime) present. The errors associated with non-compliance even increase with rising temperature.

The values given in Table 1 explain why pyrometry is to be treated critically when used without any distinct method to help distinguish thermal radiation from lamp-based radiation. Blinding of the pyrometer is a serious issue for radiation thermometry. Although neutral density filters seem to be the most obvious solution, they also decrease the sensitivity of the pyrometer and thus complicate temperature measurement due to the weak thermal radiation signal. Emissivity complicates temperature measurement even further, especially if one takes into account that it is a function of several variables, namely temperature, material, wavelength and angle. Moreover, for $\frac{1-R}{R} \gg Kd$, whereby R is the specimen's reflectivity, K its absorption coefficient and d its thickness, emissivity becomes additionally a function of sample thickness if multiple internal reflections are taken into account [50]. For ease, surfaces are often assumed to be *grey* so that emissivity is independent of the above mentioned parameters for opaque materials. However, this assumption may produce large temperature measurement uncertainties. Moreover, it is intrinsic to Planck's law that an uncertainty in emissivity causes increasing temperature measurement uncertainties for longer wavelengths and higher temperatures as shown in Figure 23.

Another source of uncertainty for radiation thermometry is thermal shadow, which is created by the sole presence of the pyrometer if lamp radiation can get reflected off it. Therefore, it is desirable to place the pyrometer outside the chamber and to view the object through a spectral window. However, if this cannot be accomplished, the shadow effect becomes a serious source for temperature measurement uncertainties. It depends strongly on the distance. The further away the pyrometer is placed from the specimen, the smaller the uncertainty which is introduced into the temperature reading by the shadow [60].

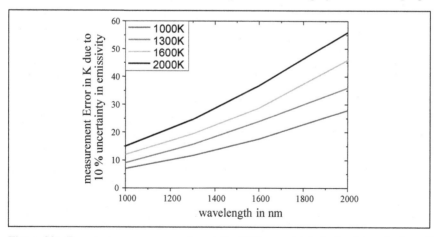

Figure 23:   Temperature measurement error for 10 % uncertainty in the determination of the sample's emissivity for wavelengths between 1000 nm and 2000 nm and temperatures of 1000 K, 1300 K, 1600 K and 2000 K (figure taken from [59]).

4 Concept of Ripple Pyrometry during Flash Lamp Annealing

4.1 Review on Temperature Measurement for Flash Lamp Annealing

Several different efforts have been undertaken to measure temperature by contactless means during RTP, for which an extensive review can be found elsewhere [3]. Some of these methods are also suitable for FLA with respect to sampling time and surface measurement. For an overview, they shall be introduced briefly in the sections that follow.

4.1.1   X-ray diffraction

If annealing occurs at constant pressure, the lattice parameter increases with rising temperature. X-ray diffraction measurements can make this fact visible by the direct relationship between the size of lattice parameters and the scattering angle according to the Bragg condition.

Upon temperature rise, an angular shift of the Bragg reflex in the diffractogram can be observed. However, in the general case, the hot sample is not uniform in emperature, but exhibits a transient thermal gradient. The paper considers the sample consisting of an unstrained (cool) substrate and a hot surface, which can be divided into a number of smaller subsurfaces, each of them at a different temperature and thus with different lattice parameters. Accordingly, this leads to a number of Bragg reflections at different angular positions. Thus, the diffractogram shows one broad Bragg peak rather than individual sharp peaks as depicted in Figure 24 [61]. Following Bragg's law of diffraction the change in the lattice parameter

$$\frac{\Delta a}{a} = -\cot[(\theta)\Delta\theta], \tag{23}$$

whereby a represents the lattice parameter and θ the Bragg angle, is directly related to the angular shift of the Bragg peak in the x-ray diffractogram. To obtain the strain distribution directly from the X-ray diffraction measurement the authors used its depth dependence as a fitting parameter (no details mentioned). This makes this method unsuitable for thermal annealing of un-

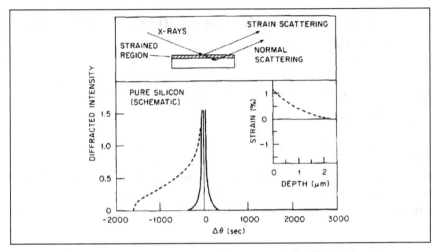

Figure 24: Illustration of the broadening of the Bragg peak in X-ray diffraction measurements due to non-uniform changes of the lattice constant (figure taken from [61]).

known material. However, the time resolution is in the ns-regime by the use of synchrotron radiation and it allows for surface measurement, which is essential for flash lamp annealing. Yet, the obtained temperature uncertainties in the order of 50 ... 75 K are too large for current industrial requirements.

4.1.2   Interference

In the following approach for temperature measurement, an infrared laser was used to irradiate the wafer backside, while its surface was annealed by a thermal plasma jet [62]. Thereby, the laser was repeatedly scattered at the front- and backside of the wafer. The resulting interference pattern was recorded (cf. (A) in Figure 25). Temperature changes during annealing can be observed by oscillations in the interference pattern (cf. (B) in Figure 25) due to a change in the optical path difference of the beams being reflected between the front- and backside of the wafer.

Outrider to this variation of temperature measurement was Murakami and his group in 1981 [63]. They investigated Silicon on Sapphire samples and measured the interference between light reflected from the Silicon surface and from the Silicon-Sapphire interface.

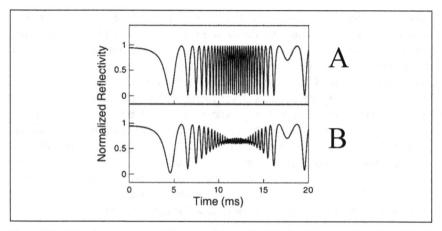

Figure 25: Interference pattern in terms of normalized reflectivity as a function of time. A: Silicon wafer at constant temperature, B: Silicon wafer with a temperature distribution (figure taken from [62]).

The refractive index has been determined offline as a function of temperature in thermal equilibrium. Thus the temperature of the sample can be simulated to match the experimental results on transient reflectivity. This, however, complicates online temperature control – an important prerequisite for flash lamp annealing – and annealing of unknown material. Though this method allows for both millisecond time resolution and a high temperature resolution with an uncertainty of 1 K.

4.1.3   Recrystallisation

Solid phase regrowth (SPR) or recrystallisation of amorphous Silicon responds immediately to changes in temperature. In fact, it is a very sensitive method due to the exponential relationship between temperature T and the velocity of regrowth

$$v \propto B \exp\left(\frac{A}{k_B T}\right) \tag{24}$$

whereby the constants A and B denote the activation energy for recrystallization and a minimum velocity, respectively. Note, that the start of spontaneous recrystallization, however, is delayed due to the missing nucleation seed at the beginning. In any case, the smaller v the more sensitive this method, which in turn shows its frontiers for MSA [64]. In fact, own unpublished experiments revealed that SPR is best used in the low temperature regime, which is unfavourable for its use during MSA or FLA. At elevated temperatures the regrowth is much faster than 1 ms and thus one cannot infer the temperature from measuring the sheath thickness as it remains unknown after which period of time the amorphous surface layer was fully regrown. However, SPR may be used as a verification and / or calibration tool for pyrometry [65]. In the year 1985, experiments were performed for spontaneous recrystallization of pre-heated amorphous Silicon on SiO_2. As indicated above, this process requires an initiation, which was accomplished by a Quantum-switched Neodymium-glass laser. Time-resolved reflectivity measurements were used to trace the propagation of the amorphous-crystalline phase front. A second laser pulse stopped the process immediately to create a sharp phase front in order to ease the determination of v.

Figure 26 shows a copy of the original experiment. It illustrates the lateral spontaneous recrystallization from amorphous to crystalline Silicon. The phase front and the thermal profile are shown. The direct relationship between recrystallization speed and temperature yields a real time thermal profile. For transient SPR the sheath thickness of the recrystallized layer may be determined by ellipsometry (non-destructive method, cf. next subsection) or by transmission electron microscopy (destructive method) [64].

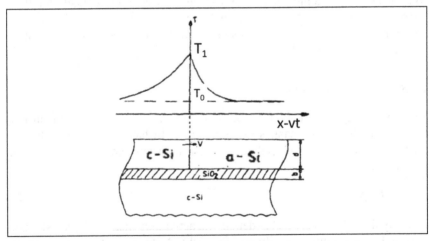

Figure 26: The lower part of the picture shows a cross-section of the wafer and the propagation of the amorphous-crystalline phase front with velocity v. The upper part shows the according temperature-time diagram with preheat temperature T_0 and peak temperature T_1, (figure taken from [64]).

A temperature uncertainty is not given, yet it can be assumed that the accuracy of temperature measurement depends on the model used for the determination of A and B. This makes this method despite its large sensitivity unsuitable for annealing of unknown material and especially for patterned wafers.

4.1.4   Ellipsometry

Single-wavelength ellipsometry is a technique that may be used to determine the optical properties of a solid surface. Their temperature dependence can be exploited for online temperature measurement [66]. Outrider to this idea in the framework of annealing was the IBM Research Division in New York for the application during plasma processing. In this paper, a Helium-Neon laser was used for carrying out ellipsometric measurements at a wavelength of 632.8 nm on a Silicon wafer surface (cf. Figure 27). Studying the reflectivity

$$R = \frac{R_p}{R_s} = \frac{\frac{E_r^p}{E_i^p}}{\frac{E_r^s}{E_i^s}} \tan\Psi \exp^{i\Delta} = f(n_{Si}, n_{ox}, X_{ox}, \phi, \lambda) \tag{25}$$

in terms of the ellipsometry parameters Δ and Ψ, where p and s refer to the s- and p- polarisation, respectively, and E indicates the electric wave vector, allowed for the extraction of a temperature reading through the knowledge of the temperature dependence of the refractive indices n and the oxide film thickness X_{OX}. θ refers to the angle of incidence λ to the wavelength of the laser [66].

In general, limitations result from the resolution of the ellipsometer, the exact knowledge of the optical constants used, the adjustment of the angle θ (+/ 0.1°C) as well as the resolution of Δ and Ψ (+/ 0.1 °C for both). These limitations themselves, however, depend on temperature. The temperature results obtained from ellipsometry measurements were compared to thermocouple readings [66]. Deviations of up to 35.4 K were observed.

For the use in online FLA, this method exhibits the strongest limitations and requires exact knowledge of the optical properties of the material, which is not favourable for industrial purposes although this method does allow for surface measurement.

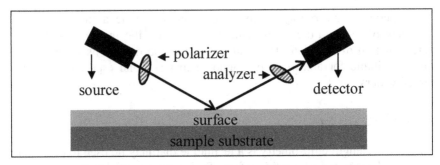

Figure 27: Schematic drawing of the set-up for ellipsometry for temperature measurement.

4.1.5   Polaradiometry

All previous methods rely on a randomly existing infrastructure on site. An upgrade for temperature measurement only would be inappropriate.

A similar technique to ellipsometry, that uses a broadband black body source instead of a laser, is polaradiometry, first published in the late 60's. However, its use for temperature measurement during FLA is still in its infancy. In the year 2010, a group from the Japanese National Institute of Advanced Industrial Science and Technology presented this method for emissivity-compensated true temperature measurement [67].

The authors modulate the source to give them an *ON* and an *OFF* radiation value $I_{ON/OFF}$ for p- and s-polarised component that is incident on the radiation detector. By comparison of the two, the thermal radiation from the sample can be extracted knowing that

$$I_{p_{OFF},s_{OFF}} = \varepsilon_{p,s}\, M_{bb} = (1 - R)M_{bb} \text{ and} \tag{26a}$$

$$I_{p_{ON},s_{ON}} = \varepsilon_{p,s}\, M_{bb} + R_{p,s}I_{source} \text{ , where} \tag{26b}$$

$$\frac{R_p}{R_s} = \frac{(I_{p_{ON}} - I_{p_{OFF}})}{(I_{s_{ON}} - I_{s_{OFF}})}. \tag{26c}$$

Repeating the measurement at a different value of I_{source} will eliminate $I_{p_{ON},s_{ON}}$ to give the black body radiation

$$M_{bb} = \frac{(I_{p_{OFF}} - I_{s_{OFF}})}{\left(1 - \frac{R_p}{R_s}\right)} \tag{26d}$$

The masking of the interfering flash light is brought about by an intelligent filter design proposed by *Vortek Industries Ltd.* at the beginning of this century [3], [4]. The hydroxyl group in water windows absorbs radiation at the wavelength of its first harmonic around 1450 nm, which is set a detection wavelength. This ensures that only radiation hits the detector that originates from the wafer. The disadvantage of which is, however, that the filter system requires extensive cooling for the strong light intensity of the flash lamps.

A temperature resolution of 8 K between temperatures of 1073 K and 1473 K is achieved for RTP control tests. The method was also applied to online measurements during FLA. However, as no reliable reference for temperature measurement during FLA exists, the sheet resistance was determined after annealing to compare to RTP measurements. The results were in good agreement. However, its complexity and the requirement of cooling lower the attractiveness of this approach for temperature measurement during FLA.

A comprehensive summary of the advantages and disadvantages of each method for temperature measurement during FLA in comparison to pyrometry can be found in [3]. Although the alternatives are promising, pyrometry is the only method that can be applied inexpensively and without a large experimental set-up.

However, as discussed in the preceding chapter, there are several limitations to the use of a pyrometer. There have been two proposals to solve this issue. One of these relies on the usage of an intelligent filter design, which enables background isolation. However, it does not compensate for emissivity changes [3], [4]. The other, in contrast, can potentially solve both issues. It makes use of the modulation of the flash and is known as ripple pyrometry. The state of the art of this technique will be closely looked at in the following section.

4.2 State of the Art [3]

To the best of the author's knowledge ripple pyrometry has not been implemented for FLA so far [4], [68]. This is mainly due to the difference in lamps between RTP and FLA. The state of the art for ripple pyrometry relies on electric modulation of circuit-driven halogen lamps in RTP [6]. Figure 28 illustrates the basic set-up for ripple pyrometry during RTP.

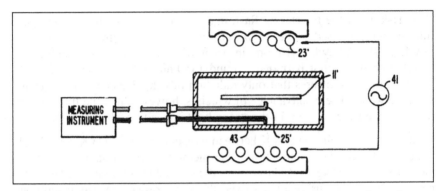

Figure 28: Sketch of an RTP annealing apparatus using ripple pyrometry for wafer temperature measurement (figure taken from [6]).

A set of detectors (no. 25', 43) view the wafer and the lamps, respectively. Accordingly, they will be referred to as *wafer* and *lamp detector* in the following.

If the lamps (no. 23') are operated by the electric circuit network a ripple (ΔI_L) of 100 Hz (so-called flicker frequency) is generated on the lamp radiation I_L due to the line frequency of 50 Hz. The signal on the lamp detector (no. 43) is displayed in Figure 29 (graph 51). After reflection off the wafer surface (no 11') ΔI_L decreases to $R \times \Delta I_L = \Delta I_W$. The wafer detector (no. 25') receives a mixed signal I_W (graph 53 in Figure 29) which is composed of the reflected flash lamp radiation off the wafer surface and thermal radiation. If one was able to measure the thermal radiation of the hot wafer separately, no ripple would be detected on it (graph 55 in Figure 29) due to its high thermal mass (during RTP). The influence of the lamps is thus identified by comparison of the modulation depths ΔI_L and ΔI_W. This allows for the determination of the reflectivity of the sample surface and thus of the share of flash lamp radiation in I_W. The latter can now be subtracted from the wafer detector signal to yield the thermal radiation alone (graph 55 in Figure 29) to give a background-corrected signal

$$E_W = I_W - \frac{\Delta I_W}{\Delta I_L} I_L. \tag{27}$$

The temperature measurement uncertainty during RTP may be less than 2 K [68].

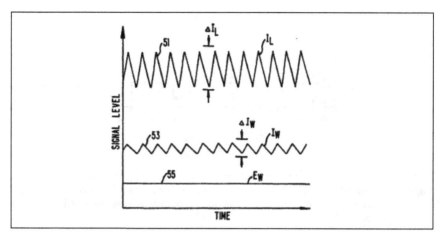

Figure 29: Sketch of the idea of ripple pyrometry in RTP. (Figure taken from [6]).
51: amplitude modulation of lamp radiation; 53: amplitude modulation of
wafer radiation (= superposition of thermal radiation and reflection of
amplitude-modulated lamp radiation; 55: thermal radiation through
background correction.

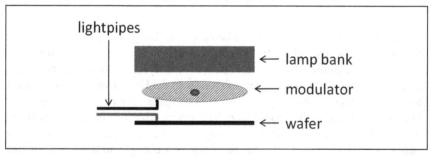

Figure 30: Schematic of a set-up containing non-circuit driven lamps like flash lamps
for using ripple pyrometry.

Although this technique is very promising, it is not applicable to FLA as the
appropriate units use the discharge of capacitor banks to drive the flash
lamps. A possible solution will be some different kind of lamp modulation
acting similarly to an optical chopper (c.f. Figure 30).

This idea has also been followed by TEXAS Instruments Inc., yet again only
for halogen lamps [69]. The author suggests an AC modulation of the heating
lamps with a frequency small enough such that the lamp filaments respond to

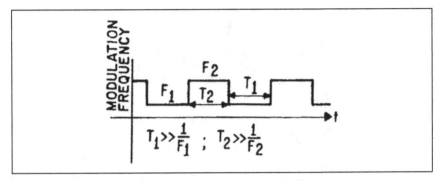

Figure 31: Time-division AC modulation scheme to eliminate lamp and hot quartz interference by taking into account the respective thermal mass (figure taken
from [*69*]).

the modulation and the pyrometer can detect it, but at the same time large
enough for the wafer radiance to stay unaffected. The reason for this behaviour is the small thermal mass of the lamp filaments compared to the wafer
material. The inventor even succeeds in eliminating the influence of heated
quartz parts for the same reason as quartz has the highest heat capacity of the
three.

Figure 31 shows the modulation scheme of the lamps. The modulation frequency F_1 is chosen to be very small (typically < 5 Hz) to extract the interference from radiation of hot quartz parts, while F2 is chosen to be much
larger (ranging from frequencies of 10 Hz to 1 kHz) in order to eliminate
lamp interference. Accordingly, if an AC voltage is applied, the piezoelectric
element oscillates at the modulation frequency, which in return gives feed-
back to the modulator, thereby locking the circuit at the modulation frequency. The device so described functions as a chopper to a beam of wafer
radiance and may be subsequently amplified by lock-in techniques, which requires several oscillation periods for reliable measurement. This makes them
– as it has been the case for electric modulation – not suitable for FLA. Yet,
investigations on the effectiveness of ripple pyrometry for true temperature
measurement showed that using a ripple control for setting annealing temperatures during RTP results in steady thermal treatment despite emissivity
changes due to phase transitions or varied doping concentrations [**70**], which
is shown in Figure 32.

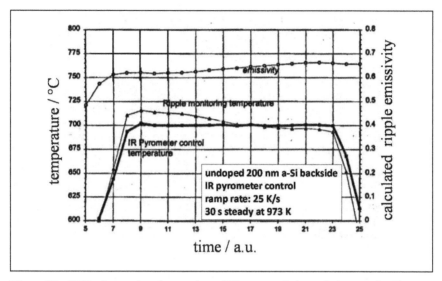

Figure 32: RTP of an undoped amorphous Silicon sample is carried out under IR pyrometer control (figure acc. to [70]).

Annealing on amorphous Silicon has been performed at a temperature ramp rate of 25 K/s up to a steady state for a time duration of 30 s. During annealing there is a transition from the amorphous state to polycrystalline material, which changes the emissivity during annealing. This results in a change in spectral power εM_{bb} impinging on the pyrometer detector as discussed in subsection 3.3.2, which leads to the pyrometer reading a different temperature than the actual sample temperature.

In a temperature-controlled annealing process a pyrometer that shows, for example, a higher temperature due to an increase in emissivity will lower the current through the halogen lamps to decrease light power output, if the change in emissivity is unknown. However, a ripple pyrometer for monitoring temperature during process control is able to identify these changes in emissivity (cf. Figure 32). Such a ripple pyrometer may consist of a conventional IR pyrometer with ripple control as explained above. These investigations have motivated the search for suitable applications of ripple pyrometry during FLA.

5 Ripple Pyrometry for Flash Lamp Annealing

5.1 Idea and Set-up

Conventional temperature measurement is associated with radiation pyrometry. However, as shown above, this becomes a difficult task to perform in lamp-based annealing devices. Therefore, offline calibration procedures are usually involved which are based on calibration points such as the melting point or the point of recrystallization. However, these methods assume that the specific heat capacity is constant in the range between the preheat and the peak temperature which is true for the heat capacity at constant volume c_V (cf. Figure 8)). For FLA, however, the heat capacity at constant pressure c_p has to be considered, which changes by about 16 % from 800 K (typical preheat temperature) to the melting point of Silicon at 1683 K (cf. Figure 8). According to Equation (4), this transforms into an uncertainty in temperature measurement up to 120 K.

For these reasons, online methods are needed for true temperature measurement. For the short time-regime during FLA, non-contact techniques are required for temperature measurement. Besides the possibility of a genuine filter design, a method known as ripple pyrometry based on online reflectance measurements has been proven successful for a number of applications.

The idea for ripple pyrometry during FLA is illustrated in Figure 33. Thermal radiation from the wafer is superimposed with modulated flash light, which has been reflected off the wafer surface. If it was possible to detect both signals individually, the ripple on the modulated flash pulse would show to be of equal size as the ripple on the combined signal. Therefore, if the ripple size ΔI_L of the incoming flash pulse I_L (not shown) is known, the reflectivity of the sample surface, off which the flash is reflected, can be obtained computing the ratio of ΔI_W and ΔI_L. The mathematical description, set-up and proof of concept of the mechanical oscillator as it has been used for temperature measurement during FLA within the context of the present work will be studied in detail later beginning in section 5.3. In the sections that follow solutions to implement ripple pyrometry into FLA will be presented.

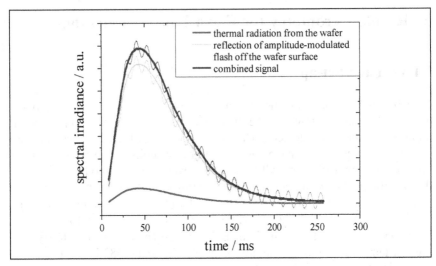

Figure 33:   Schematic of ripple pyrometry for FLA. The thermal radiation from the wafer is superimposed by the reflected flash light carrying an amplitude modulation (ripple). The pyrometer sees the superposition of both signals.

5.1.1   Intrinsic Ripple

The flash pulse forming *LC* resonator of the flash lamp annealer, which is shown in Figure 34, is used to create millisecond flash pulses of various shapes and sizes. As a consequence of self-induction in the coils due to the decreasing current after the capacitor has been discharged a ripple is created on the flash lamp light output, which may be referred to as intrinsic ripple.

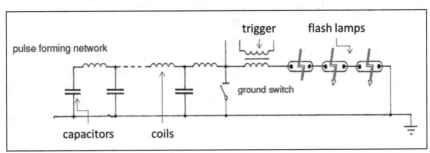

Figure 34:   Flash pulse forming LC resonator network.

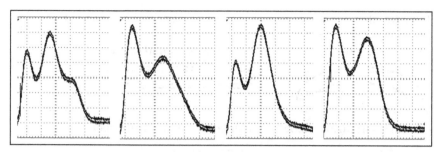

Figure 35:   Ripple creation on a flash pulse by suitable coil-capacitor-arrangements. The x-coordinate division is 5 ms in time, the y-coordinate shows the intensity in arbitrary units. The coils (with an inductivity in units of mH) and capacitors (with a capacity in units of mF) needed to create the flash pulse have been arranged as follows from left to right:
A: 0.41 mH – 1 mF – 0.81 mH – 2 mF – 0.81 mH – 2 mF
B: 0.34 mH – 2mF – 0.68 mH – 2 mF – 0.68 mH – 2 mF
C: 0.28 mH – 1 mF – 0.57 mH – 2 mF – 0.28 mH – 2 mF
D: 0.28 mH – 2 mF – 0.56 mH – 2 mF – 0.28 mH – 2 mF

Figure 35 shows the intrinsic ripple with an effective modulation frequency of $\frac{1}{2t_A}$ for varying coil-capacitor-arrangements. However, as this way of imposing a ripple on the flash light is limited to a few modulations for the given technical possibilities it would not be sufficient for accurate determination of the sample reflectivity and thus temperature measurement in the sense of this work.

An alternative method for creating amplitude modulation in order to perform ripple pyrometry for FLA may be fast switching of the capacitor bank which will, however, still be governed by the discharge time constant. An increased number of coils in the LC resonator circuit retards the discharge and lengthens the flash pulse width. An alternating arrangement of coils and capacitors leads to ripple formation as illustrated above. It may be used to deduce the sample reflectivity using the method ripple pyrometry as explained above, which will be considered in more detail in the next section.

5.1.2   Quartz Oscillator

Quartz oscillators are specially cut SiO_2 piezoelectric crystals that show a behaviour depending on the direction of cut. A comprehensive summary on quartz oscillators can be found in [71].

Table 2:   Overview on the deformation modes for quartz oscillators (acc. to [71]).

Flexural mode	Length extensional mode
Areal shear strain mode	**Thickness shear strain mode** Symmetric: Twisted: Antisymmetric:

Piezoelectric (Greek: πιέζειν – to press) crystals respond mechanical deformation by a change in polarisation creating an electric field (direct piezoelectric effect). The inverse phenomenon is known as indirect piezoelectric effect. Alternating current causes oscillations at set frequencies which can be described by Equation (1) for an LC-resonator. The damping of the resonant circuit τ_D is due to the viscosity of the material, the finite size of the quartz oscillator as well as due to the electric contacts. It directly relates to the quality factor

$$Q = \frac{\pi f_{res}}{\tau_D} \tag{28}$$

considering the resonant frequency f_{res}. The oscillation mode of the quartz depends on its cut and is summarised in Table 2.

Distinct properties are attributed to the crystallographic axes of a single quartz crystal. Figure 36 shows a schematic of the atomic arrangement in the crystal cell. Upon electric irritation of the x-axis as defined in Figure 36 the corresponding y-axis deforms mechanically. A force along the "mechanical" y-axis results in an increased charge separation along the "electric" x-axis and therefore in an increase of the dipole moment along the x-axis and vice versa.

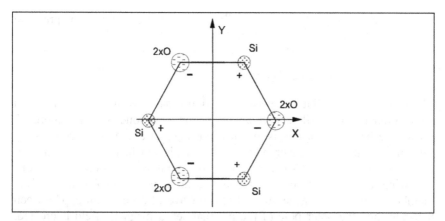

Figure 36:   Schematic drawing of the hexagonal quartz cell. The leading signs indicate the respective electronegativity; O = Oxygen, Si = Silicon (figure taken from [71]).

The effects described are reversible. However, the frequency response itself is a function of both temperature and time. The so-called ageing effect, which is a combined consequence of material aspects, crystal growth and fabrication, causes a long time shift of the resonance frequency f

$$\frac{\Delta f}{f} \propto \exp\left(\frac{-E}{k_B T}\right) \ln t \tag{29}$$

as a function of activation energy E and time t [71].

The amplitude modulation of the flash radiation may be accomplished on two different routes. One possibility is brought about by using the so-called SC-cut which exhibits oscillations in the length extensional mode (cf. Table 2). The resulting modulation of the optical path length leads to an amplitude modulation following an oscillating degree of absorbed incident light. Alternatively, if a gridding having a grating constant $g >> \lambda_{incident\ light}$, is deposited onto the quartz oscillator with an AT-cut, which oscillates in the thickness shear strain mode, similarly to the SC-cut, an amplitude modulation of the incident light can be expected. However, the displacement of the oscillator in the μm-regime is a crucial limitation, because it requires a grating constant g and a detector size of the same order of magnitude. If, however, a displacement in the mm-regime is aimed at, the quartz oscillator may not withstand the enormous shear stress and the voltage required of several 100 V will cause laborious piezoelectric actuation. Based on the same idea, but using a handier technology, mechanical oscillators are more suitable for practical use.

5.1.3   Mechanical Oscillator

The mechanical oscillator consists of a sinusoidal moving system based on conventional voice coil technology as found in a loudspeaker. Figure 37 shows the basic layout consisting of a pot magnet and a voice coil. In the case of an actual loudspeaker, a thin membrane is attached to the moving coil of the system. The AC-driven voice coil exhibits harmonic oscillations following the Lorentz force. Consequently, the harmonic displacement of the membrane leads to the generation of sound waves. Inside the air gap between the inner and the outer pole of the pot magnet, a magnetic field is created. The gap ought to be as small as possible to increase the magnetic field strength. The displacement of the voice coil needs to be controlled such that it does not exceed the height of the air gap. In such a case the leakage magnetic field would destroy the harmonic oscillation [72].

In contrast to loudspeakers, voice coils are designed to move heavier objects. Solving the differential equation for forced and damped oscillations by the damping factor δ, the maximum Lorentz force F_0

$$F_0 = s_0 m \sqrt{(\omega_0{}^2 - \omega^2)^2 + 4\delta^2 \omega^2}, \tag{30}$$

that is required to move an object of mass m by a distance s_0, whereby $s = s_0 \sin(\omega t + \varphi)$, $F = F_0 \sin(\omega t)$, the eigenfrequency ω_0 is expressed by $\sqrt{\frac{D}{m}}$ with D specifying the modulus of silence and δ can be calculated by $\frac{R_{oh}}{2L}$, where R_{oh} gives the ohmic resistance and L the inductivity of the voice coil [73].

$$R_{oh} = \sqrt{(Z^2 - \omega^2 L^2)} \tag{31}$$

can be obtained from the frequency dependent impedance Z of the system. For low frequency operation, however, ωL approaches *zero* [74]. Nowadays commercially available high-performance voice coils can generate forces up to 1 kN.

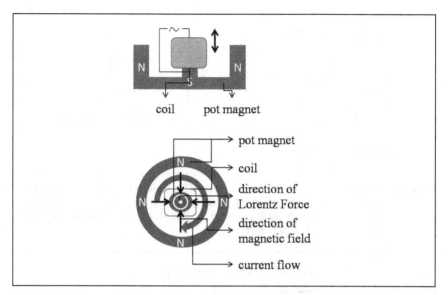

Figure 37: Schematical set-up of a voice coil. Top: Illustration of the displacement of the voice coil due to the Lorentz force. Bottom: Top-view of the voice coil.

Figure 38 is a schematic diagram of a mechanical oscillator consisting of a voice coil and a grooved quartz plate in an oscillating system, which is illustrated by the mechanical spring, for ripple pyrometry during FLA. A suitable choice of thespring assures oscillation near the eigenfrequency and at low energy. The voice coil moves in a sinusoidal manner a quartz plate on or into which a grid is deposited for amplitude modulation. This system replaces the modulator in Figure 30.

The grid can be either absorbing or reflective, but it is important to note that the grating constant of the grid g, the displacement s of the voice coil and thus the displacement of the grid and the dimensions of the active area of the detector should be at the same order of magnitude in size, which has a considerable effect on the modulation depth / ripple size of the lamp radiation. However, the detected modulation frequency may be a multiple of the set frequency of the voice coil which is indicated in Figure 39. If the displacement s of the mechanical oscillator under the flash lamps is a multiple of the grating constant g, the detected ripple frequency will also be a multiple of the set oscillation frequency. This allows for smaller forces, while the latter remains unchanged.

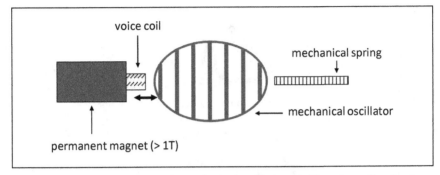

Figure 38: Schematical set-up of a mechanical oscillator aided by voice coil technology. The voice coil on the left-hand side continuously pushes a grooved quartz plate and the mechanical spring on the right-hand side serves for continuous oscillation of the quartz plate.

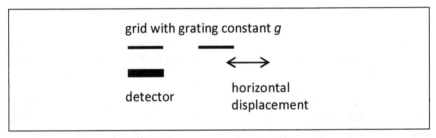

Figure 39: Sketch of the mechanical oscillator during FLA. A suitable choice of size of the grid, the detector and the horizontal displacement allows for a multiplication of the ripple frequency w.r.t. the oscillation frequency of the voice coil.

5.2 Methods

To investigate temperature measurements for millisecond annealing, the following methods have been used:

5.2.1   Temperature Simulation

A one-dimensional simulation of the thermal evolution in space and time is done by two programmes developed by Dr. Lars Rebohle at the *Helmholtz-Zentrum Dresden-Rossendorf* and Dr. Mark Smith at the *University of Cambridge*. They are based on the one-dimensional heat flow equation

$$\frac{\partial}{\partial t}(\rho h) = \frac{\partial}{\partial x}\left(\lambda \frac{\partial T}{\partial x}\right) + S \tag{32}$$

assuming a constant surface temperature T, where ρ represents the material density, h is the enthalpy, λ denotes the thermal conductivity and the source S refers to the incident flash energy density and losses due to convection or thermal radiation [75]. The boundary conditions are set to be adiabatic, i.e. there is no heat flow except for the one given by S. The incoming flash needs to be specified by the user in terms of the total energy density deposited onto the wafer, the flash spectrum and the normalised temporal flash pulse. Moreover, the program requires the input of the complex refraction index. The real part n gives information on the phase speed inside the material and the imaginary part k is needed to calculate the absorption coefficient for each point of the flash spectrum knowing k of the ambient atmosphere. The enthalpy is given through the input of the heat capacity

$$c_p = \frac{1}{\rho}\left|\frac{\partial h}{\partial T}\right|_p \tag{33}$$

at constant pressure and the material density ρ. Losses due to gas convection at the surface

$$Q_{conv} = \beta A(T - T_{env}) \tag{34a}$$

and radiative losses

$$Q_{rad} = \varepsilon \sigma A(T^4 - T_{env}^4) \tag{34b}$$

are also taken into account, where β denotes the heat transfer coefficient, T_{env} the ambient temperature, ε the emissivity and σ the Stefan-Boltzmann constant.

The programme enables the user to specify a thin layer system on top of the sample substrate. In addition, the simulation time as well as the time and spatial resolution can be chosen. The preheat temperature, which is required, sets the initial temperature of the sample.

5.2.2   Secondary Ion Mass Spectrometry [76]

Shallow Boron-doped Silicon samples have been annealed and the temperature during annealing has been recorded. Afterwards the concentration profile of the diffused dopants has been measured by SIMS and the

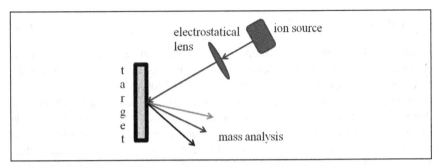

Figure 40: Schematic diagram showing the basic principle of secondary ion mass spectroscopy.

measurements were compared to simulations on the dopant diffusion based on the temperature-time-profile determined. A match between SIMS simulations and measurements would verify the temperature measurement during annealing using the technique of ripple pyrometry.

By bombarding a target with highly energetic particles (primary ions with mass M_1) neutral and ionized target atoms are emitted from the surface (secondary ions with mass M_2).

The energy transfer factor

$$\gamma = \frac{4M_1M_2}{(M_1+M_2)^2} \tag{35}$$

contributes to the sputter rate S_p. Figure 40 shows a schematical diagram of the set-up and the function of SIMS. An ion source emits primary ions which are collimated by an electrostatic lens and directed onto the target surface. By this process secondary ions are emitted, which are then detected separately by their mass.

The sputter rate S_p of the target is a function of the atomic mass of the incident ions, of their energy distribution, of the matrix of the target as well as of the target atomic mass. Upon bombardment several processes may occur depending on the incident energy. For low incident energy of a few eV single collisions are more probable, whereas for energies between keV and MeV collision cascades will take place. In general, the elements to be detected are neutral while they remain in the target matrix. Initiated by ion bombardment, collision ionisation will occur which is primarily relevant for surface atoms. Elements from beneath the surface may undergo electron capture or emission, or they may self-ionise from an excited state during their travel towards the surface. However, SIMS is usually performed in the low energy regime without deep penetration of the incident ions.

The flux of secondary ions

$$I_S = I_p S_p c \, f'$$ (36)

relates to the flux of primary ions I_p, the sputter rate S_p and the concentration c of the species to be analysed, where f' is a device specific quantity. The emitted secondary ions are extracted from the target surface by a strong electric field and are accelerated to one and the same velocity before entering the analyser. Two commonly used analysing systems are mass spectrometry and time of flight analysers.

Mass spectrometry makes use of the fact that the radius

$$r = \frac{mv}{qB}$$ (37)

of deflection inside a magnetic field with magnetic flux density B is a function of the mass to charge ratio $\frac{m}{q}$ for equal velocity v (i.e. equal kinetic energy). Neutral elements, however, cannot be detected as they would pass the rear magnets of the spectrometer unhindered. More frequently used time of flight analysers make use of the fact that the secondary ions possess equal kinetic energy. Thus, light ionised species are faster than heavier ones. This method allows for higher sensitivity and is thus, suitable for detecting low concentrations.

SIMS can be run in two modes – lateral and depth profiling, whereby the latter is the most important one for material science to investigate the chemical composition of semiconductors. Elements at a concentration of less than 1 ppm can be detected into a depth down to a few μm. More detailed information can be found elsewhere [76].

5.2.3   Reflectometry

Reflectometry is a technique that measures the spectral reflectivity of solid samples. The results can be transformed into absorptivity readings for zero transmission. In the present work a UV-Vis-NIR reflectometer from *Shimadzu* has been used with a photomultiplier for the UV-Vis regime down to a

wavelength of 165 nm as well as an InGaAs and a PbS detector for high
sensitivity in the NIR regime (model: SolidSpec 3700 DUV).

5.2.4   Raytracing

For the purpose of this research, a raytracing programme has been developed
using the system-design platform *LabVIEW*^{TM} from *National Instruments
Corporation*. The program analyses the method of ripple pyrometry for light
pulses based on trigonometry. The detector signal is simulated for various
positiofns, angles and sizes of the detector. All input parameters – size,
position and shape of reflectors; size and position of lamps and wafer as well
as the specifications for the mechanical oscillator – can be chosen according
to the needs of the user. Rays that possess a common point of intersection
with the lamp or wafer detector are counted.

The configuration of the simulation is shown in Figure 41. The mechanical
oscillator, (here: grid) as described above, is positioned between a set of 8
flash lamps and the wafer. Between the grid and the wafer, the lamp detector
and the wafer detector can be found. The grid serves as a barrier to the rays
coming from the lamps. The sample reflectivity is held constant at *one*. This
is a reasonable assumption according to the temperature dependency of the
reflectivity (cf. Figure 18). The ripple formation for various settings is investigated to verify the suitability of ripple pyrometry for FLA processes.
Details will be given in the course of the following sections.

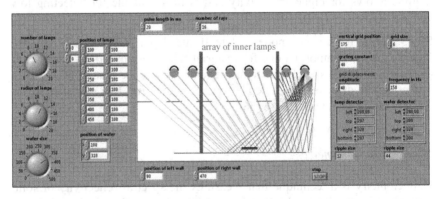

Figure 41: Raytracing simulation for ripple pyrometry during FLA. Reflection off the
grid, off the chamber walls and off the wafer is taken into account. The two
bars at the bottom illustrate the wafer detector and the lamp detector.

5.3   Technical Realisation

The results described in the following chapters were carried out in an FLA chamber as shown schematically in Figure 42. Flash lamps by the *Heraeus Holding GmbH* were used to generate pulses with a time duration of 20 ms, 40 ms and 80 ms. Table 3 summarizes the technical specifications. The mechanical oscillator, which is displayed in red in Figure 42 is composed of an *indEAS©* voice coil. The two quartz plates serve as protection to both the lamps and the wafer.

Table 3:   Technical specifications of the flash lamps used and to generate flash pulses with a time duration of 20 ms, 40 ms and 80 ms.

Flash pulse duration	Required capacitance and inductance for the LC circuit	General features of the flash lamps
20 ms	$L = 66$ mH, $C = 6$ mF	Length: 280 mm
40 ms	$L = 132$ mH, $C = 12$ mF	Diameter: 13 mm Glass envelope:
80 ms	$L = 264$ mH, $C = 24$ mF	fused quartz Fill gas: Xenon at 600 hPa

For the online determination of the wafer reflectivity R at RT broadband detectors (photodiodes) with a wavelength regime between 350 nm and 1100 nm were used, while for the experiments at elevated temperatures an *IMPAC* (*LumaSense Technologies*) pyrometer with a wavelength detection regime between 800 nm and 1100 nm for temperatures > 873 K measured I_W and R was taken from reflectometry measurements. Due to technical limitations at elevated temperatures R could not be determined online. The broadband detectors view the lamps through the grooved quartz plate and the wafer, respectively, as indicated in Figure 42. An unstopped line of sight for the pyrometer is assured by a hole in the chamber ceiling.

The complete set-up for the mechanical oscillator consisting of the voice coil, an U-head and the quartz wafer is shown in Figure 43 for clarity. It consists of a grooved quartz plate with an absorbing grid and a grating constant of 5 mm, a stainless steel U-head and a counter support, a push rod and a voice coil in an Aluminium chassis. The spring from Figure 38 is made redundant as the voice coil takes over its task. The voice coil is attached to the grooved quartz plate by a U-head carrying the wafer such that the voice

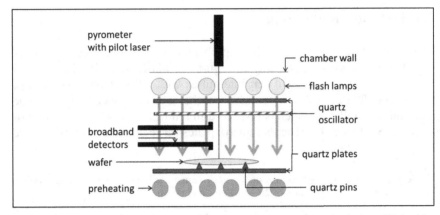

Figure 42: Experimental set-up for temperature measurement during FLA. An unstopped line of sight for the pyrometer is assured by a hole in the chamber ceiling. Its adjustment is facilitated by the pilot laser. The broadband detectors are used for reflectivity measurement.

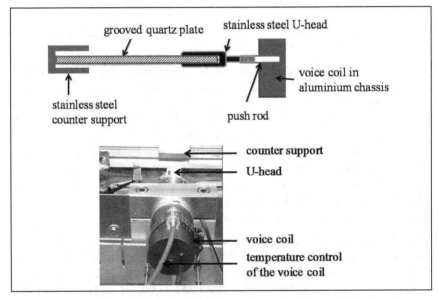

Figure 43: Experimental set-up (top) and picture (bottom) of the mechanical oscillator.

coil, which does not withstand ambient temperatures above 403 K, can be placed outside the chamber for cooling. The counter support is flooded by air to support the quartz plate at minimum friction. A push rod moves the quartz plate in a harmonic manner due to the sinusoidal changing Lorentz force acting on the AC-driven voice coil inside a magnetic field. When the flash light passes the mechanical oscillator, which moves with a time constant t, the light pulse with a total height I_L that hits the wafer is amplitude-modulated at a frequency $f_{VC} = \frac{1}{t}$ with a ripple size ΔI_L as seen by the lamp detector. This process is illustrated in Figure 44.

Upon reflection off the wafer surface, the ripple size will decrease to ΔI_W depending on the reflectivity of the sample. The detector facing the wafer (subsequently called *wafer detector*) senses a signal of a decreased height I_W. I_W is composed of the thermal radiation from the wafer and the reflected flash pulse, which is also decreased by the reflectivity of the sample compared to the value of I_L. This relationship can be summarised as follows whereby the initial parameters of the system are at the left hand side of the arrow, the adjacent path of the light is denoted above it and the right-hand side of the arrow shows the final read-out parameters of the system:

$$I_L \xrightarrow{\substack{passing\ the \\ oscillator}} I_L, \Delta I_L \xrightarrow{\substack{reflection \\ off\ wafer \\ surface}} I_W, \Delta I_W,$$

The influence of the amplitude-modulated annealing radiation from the flash lamps on the thermal radiation of the wafer surface will be investigated in section 5.5.

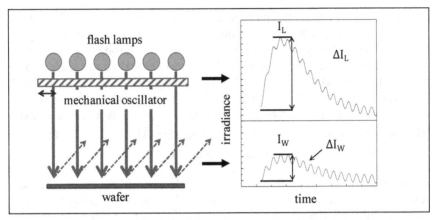

Figure 44: Amplitude modulation of an incident flash by means of a mechanical oscillator. Right: Modulated flash pulse shown for the incident case (top) and the reflected one (bottom).

5.4 Amplitude Modulation of Flash Pulses

The ripple size of the amplitude modulation of the lamp radiation

$$\Delta I_L \propto F(t) D_L \left| \frac{g}{l_{det}} \right|_{if\ l_{det}+s>g} \left(\frac{1-\tau}{2} \right) sin(\frac{s}{g} 2\pi f_{VC} t + \varphi) \tag{38}$$

can be mathematically described by a sinusoidal function, where $F(t)$ describes the unmodulated flash lamp pulse, D_L gives the responsivity of the lamp detector, l_{det} the detector size, τ the transmissivity of the grid, s the displacement of the mechanical oscillator and thus of the grid, g gives the grating constant, f_{VC} the frequency of the voice coil movement and φ represents a phase shift with respect to the progressing time t. If, however, $l_{det} + s < g$, flash lamp radiation may be lost, but no ripple will be created. The sensitivity of ΔI_L for varying parameters in Equation (38) is shown in Figure 45. The linear dependence of the ripple size on the unmodulated flash pulse, i.e. on the incident intensity as a function of time, indicates appreciable influence on the temperature measurement for a change in flash lamp radiation and supports the necessity of an online-control during FLA. The detector's responsivity, however, is generally known. An increase in the uncertainty of the knowledge of the transmissivity τ of the grid decreases the

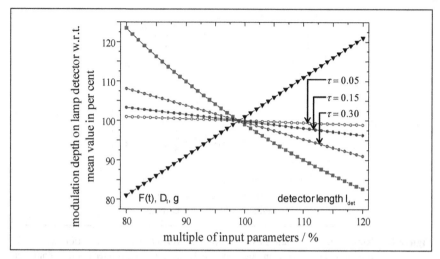

Figure 45: Sensitivity of the parameters in Equation (38) on the modulation depth ΔI_L as seen by the lamp detector.

ripple size. However, with small nominal values of τ the sensitivity of ΔI_L on τ remains small (cf. Figure 45).

For the purpose of the simulation of ripple pyrometry during FLA for various device geometries and designs, a raytracing programme has been developed during the course of this work using the *LabView®* development platform from *National Instruments AG*. For details on the programme please see subsection 5.2.4. For ease the simulation neglects preheating. This is because the preheating also contributes considerably to the background radiation [59]. To avoid the interference of the preheat radiation, it also needs to be amplitude-modulated and evaluated. To avoid this issue during experimentation, the preheating has been turned off before the flash.

For the presented technique of ripple pyrometry during millisecond annealing the positioning among the grid, the detector and the lamps as well as the value of the grating constant compared to the detector size and the displacement of the grid have been studied to investigate their influence on the ripple size. Upon varying the vertical position of the detector with respect to the grid a reduction in ripple size has been observed (see Figure 46). This behaviour follows directly from the decrease of $|F(t)|$ (cf. Equation (38)) according to the distance law for point sources. In fact, the lamp bank may

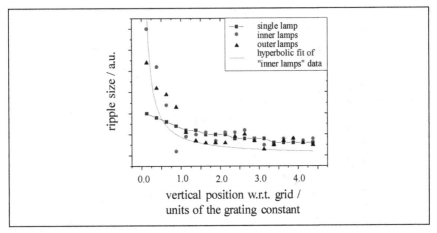

Figure 46: Ripple size simulation as a function of vertical detector position w.r.t. grid position. The inner lamps and accordingly the outer lamps refer to the definition in Figure 41.

be regarded as an assembly of point sources in close vicinity, whereas at large distances it may be compared to plane wave illumination. Consequently, at large distances from the grid, the ripple on the detector signal changes only slowly, while in the proximity of the grid the ripple size decreases significantly with increasing distance. This effect gets stronger if the detector is placed at a mid-position beneath the lamp bank due to the modulated light of the neighbouring lamps.

For comparison, Figure 46 also shows the decrease of the ripple for a single lamp, i.e. a point source. Consequently, if the detector is placed below the outer lamps of the lamp bank (cf. Figure 46), the decrease in ripple size is less steep, but can be compared to the ripple size for a single lamp. If the detector is placed below the inner lamps, however, the decrease in ripple size is steeper. The data have been fit with a first-order hyperbola due to the distance law for point sources, which changes to $\frac{1}{distance}$ for two dimensions.

The proportionality in Equation (38) is further described by the angle of the detector with respect to the grid. This is due to an overall loss in intensity incident on the detector and due to stray radiation from the neighbouring lamps. The decrease in ripple size may be fit by a sinusoidal function due to the trigonometric relationship between angle and the projected detector

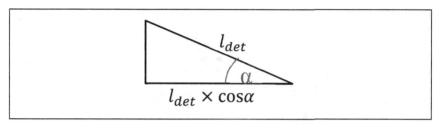

Figure 47: Sketch showing the relationship between the detection length l_{det} and its projection.

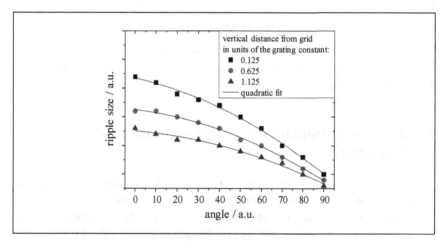

Figure 48: Ripple size as a function of angle.

size l_{det}, namely the cathetus. For better understanding, this relationship is schematically drawn out in Figure 47.

The simulation of the ripple size as a function of the angle is displayed in Figure 48. The ripple formation has been also investigated with respect to the sizes of the detector l_{det}, the grating constant g and the displacement of the grid s. The prerequisite for ripple formation is that the sum of s and l_{det} should yield a value larger than g. If this is fulfilled, the ripple size relative to its absolute value decreases with increasing l_{det} as illustrated in Figure 49. This is because the size of s and g decreases with respect to the increasing detector size (l_{det}). Thus, the relative change in intensity during one oscillation of the mechanical oscillator decreases, which minimizes the ripple size.

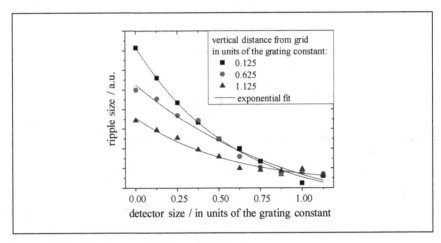

Figure 49:   Ripple size as a function of detector size.

5.5 Analysis of Amplitude Modulation during Flash Lamp Annealing

Figure 50 shows the interior of the voice coil used. Two strong Neodymium-Iron-Boron magnets with a magnetic flux density of 1.26 T induce a magnetic field into the air gap between them. The copper coil, which is placed between the chassis and the coil body, is forced by the Lorentz force to move the plunger forward and backward if an AC current is applied.

Figure 51 shows the ripple formation for flash pulse times of 80 ms at FWHM. The unmodulated flash pulse was recorded prior to the measurement. Using Equation (38) the obtained amplitude modulation could be verified. It is important to note, that the varying ripple size ΔI_L in Figure 51 is a consequence of the direct proportionality between $F(t)$ and ΔI_L in Equation (38).

The detector was chosen such that its size merges the grating constant of the grid, i.e. $l_{det} = g$. The transmissivity of the grid is < 10%. This means that a modulation depth of ca. 90 % ought to be expected if stray radiation played no role. However, Figure 51 clearly demonstrates that the modulation depth is many times smaller. This large fraction of unmodulated stray radiation will be of importance again in chapter 6.

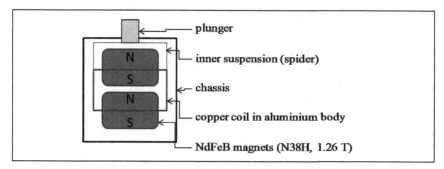

Figure 50:   Interior set-up of the voice coil used in the presented work.

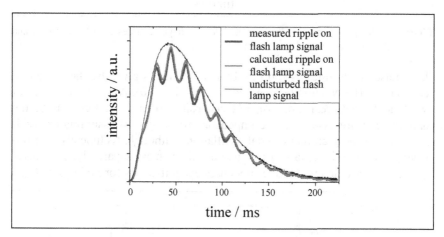

Figure 51:   Ripple formation on a flash pulse with a time of 80 ms at FWHM. The as-measured ripple is well described by applying Equation (38). The undisturbed flash lamp signal cannot be determined simultaneously. It is included in arbitrary units for comparison only.

Ripple formation for flash pulses with times of 20 ms, 40 ms and 80 ms at FWHM is shown in Figure 52. For shorter pulses a higher modulation frequency is preferable to decrease the measurement uncertainty for the ripple size and thus for the reflectivity of the sample. The pre-existing technical conditions in the annealing machine and the mechanical mounting suspension of the voice coil for the present work, however, only allowed for frequencies below 40 Hz. In contrast, Figure 53 illustrates that the technical potential for amplitude modulation up to single-digit kHz ripples and thus for

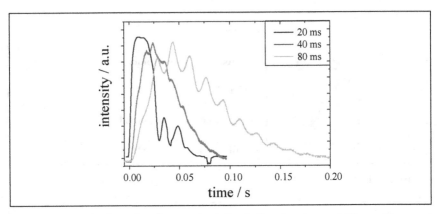

Figure 52: Ripple formation demonstrated for flash pulse times of 20 ms, 40 ms and 80 ms at FWHM.

flash pulses with times as short as 1ms at FWHM is given by the voice coil technology. The calculation is based on Equation (30). A maximum force of 280 N and a DC resistance of 3 Ω according to the specifications of the manufacturer are assumed. The impedance was obtained assuming an oscillator mass of 60 g and sinusoidal oscillation. Although, technically, the impedance needs to be obtained for each oscillator mass separately, for convenience, it is assumed that it does not change significantly for higher masses.

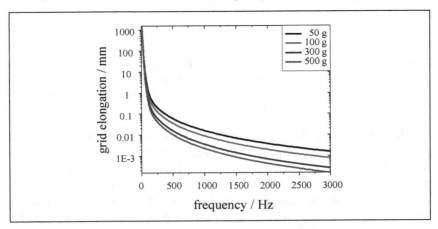

Figure 53: Maximum displacement of the mechanical oscillator and thus of the grid as a function of frequency and oscillator mass. The calculation is based on Equation (30).

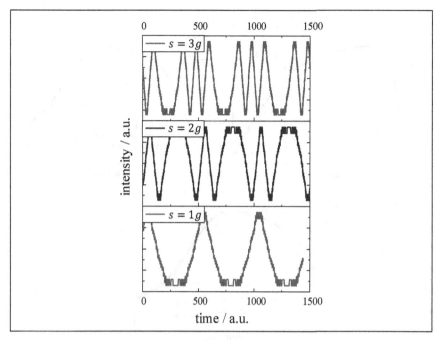

Figure 54:   Simulation of the increase of the number of ripples during a flash pulse as a function of the ratio between grid displacement s and grating constant g.

Moreover, the optical means of creating the ripple allows for multiplication of the frequency, which is set by the mechanical oscillation of the voice coil plunger. As a consequence, Figure 54 shows the ripple frequency on the modulated flashpulse being a multiple integer of the oscillation frequency of the voice coil according to the ratio $\frac{s}{g}$ as defined in Equation (38).

During RTP, pulses in the order of seconds are used, which result in complete heating of the entire wafer up to the desired temperature. Due to the large thermal mass of the entire wafer an amplitude modulation of the thermal radiation during ripple pyrometry for RTP can be excluded [6], [69]. For millisecond processes, however, a thermal gradient exists through the bulk of the wafer for light pulses with a time duration below 20 ms. Near-surface layers, which are at the process temperature possess a smaller thermal mass than the much colder bulk and may therefore respond faster to variations of the incoming flash pulse.

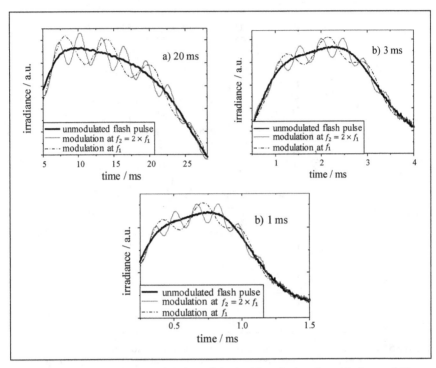

Figure 55: a–c: Amplitude modulation of the incident flash pulse with times of 20 ms, 3 ms and 1 ms at FWHM at two different frequencies: single and double.

Simulations have been performed to investigate the wafer modulation for FLA. For this purpose, two modulation modes were chosen, each applied to flash pulse times of 1 ms, 3 ms and 20 ms at FWHM. In Figure 55 the unmodulated and the modulated flash pulse with single and double modulation frequency according to the flash pulse time at FWHM are shown. Using the simulation programme described in subsection 5.2.1, the thermal profile during the displayed flash pulses is obtained with a maximum temperature of 1500 K.

Thereafter the temperature difference between the non-modulated and the modulated flash pulse was calculated (Figure 56). Although each modulation mode shows an impact on the temperature readings, the resulting uncertainty in temperature measurement is < 0.70 % for single and < 0.35 % for double modulation frequency. For decreasing flash pulse times (cf. Figure 21) the

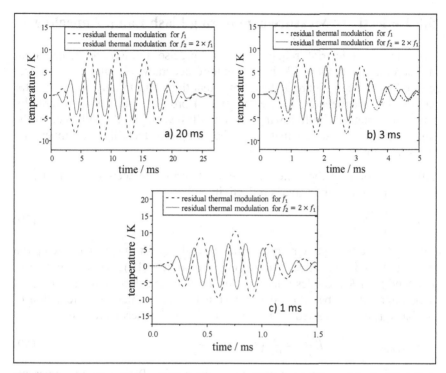

Figure 56: a–c: Temperature deviation upon amplitude modulation of the incident flash pulse with times of 20 ms, 3 ms and 1 ms at FWHM (cf. flash pulses shown in in Figure 55). Peak temperature: 1500 K.

mass of heated material and thus thermal mass decreases and hence the lower limit frequency, which is necessary such that the wafer can respond to the modulation, decreases as well. Therefore, a stronger influence on flash pulses with a time of 3 ms and 1 ms at FWHM had been expected. However, similar results in terms of temperature uncertainty were obtained (cf. Figure 56). It seems that there is a balance between the amount of thermal mass and the modulation frequency. Clearly, the frequency of the amplitude modulation for shorter annealing pulses needs to be accordingly increased. The further this frequency is increased, the less the wafer can respond.

5.6 Reflectivity Measurement during Flash Lamp Annealing

The current industrial requirements for precise millisecond temperature measurement call for 7.5 K in temperature accuracy, repeatability and precision [3]. Though Figure 56 has shown that this aim can only be reached for large ripple frequencies, Figures 53 and 54 proof that this is achievable by ripple pyrometry. The following section will investigate the influence of the uncertainty in the determination of the sample reflectivity on temperature measurement.

After reflection off the wafer surface, as described in section 5.3, the ripple ΔI is decreased by the wafer's reflectivity

$$R = \frac{\Delta I_W}{D_W} \frac{D_L}{\Delta I_L} \tag{39}$$

relative to the respective responsivities D_L and D_W for the detector facing the lamps (L) and the other one facing the wafer (W), respectively. The knowledge of R provides a measure of the reflected flash light signal that is superimposed on the wafer detector signal. This enables the calculation of the background-corrected areal power density of the thermal radiation

$$I_{thermal} = \frac{I_W}{D_W} - R \frac{I_L}{D_L} = \frac{1}{D_W}\left(I_W - I_L \frac{\Delta I_W}{\Delta I_L}\right), \tag{40}$$

from which the temperature may be obtained using Planck's law. Apparently, the responsivity D_L of the lamp detector does not influence the measurement as long as it does not change during amplitude modulation. Based on Planck's law (cf. Equation (19) Figure 57 analyses the sensitivity of the thermal radiation $I_{thermal}$ on R. The uncertainty in R decreases for increasing energy density of the incident flash due to the increasing ratio of thermal radiation as a consequence of the wafer temperature T to reflected flash radiation with energy deposited into the material to raise its temperature. This can be explained by the fast growing fourth-order polynomial of thermal radiation in contrast to the linear increase of temperature with input energy.

The uncertainty of determining the value of R clearly is a function of the accuracy in finding the ripple sizes of the lamp and the wafer detector signal ΔI_L and ΔI_W, respectively.

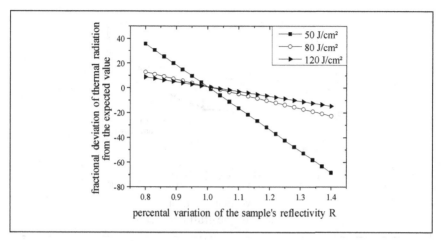

Figure 57:   Sensitivity of thermal radiation I_T on the sample's reflectivity R for different energy densities of the incident flash light.

The lower limit uncertainty

$$\sigma_R = \frac{\sigma_P}{\Delta I_L} \tag{41}$$

in determining the reflectivity R is given by the sensitivity of the pyrometer σ_P and the ripple size ΔI_L. The signal-to-background ratio is mainly governed by the ratio of thermal radiation compared to the energy density, which is reflected off the wafer surface, necessary to obtain the desired temperature increase of the sample surface. An overview for flash pulse times of 20 ms at FWHM and temperatures between 1300 K and 2000 K can be found in Table 1. Although undesirable with respect to the weak signal from thermal radiation the larger the background I_L, the larger ΔI_L for constant transmissivity of the grid and thus σ_R decreases. This means that large background benefits the accuracy of online determination of the reflectivity R. However, for the overall uncertainty in T the reproducibility of R needs to be taken into serious consideration (cf. Figures 57 and 59).

Using the raytracing programme described in subsection 5.2.4, the ripple size of the lamp detector and the wafer detector was compared to study the as-measured reflectivity results of a sample with a model reflectivity of *one* (cf. Figure 58). The most important requirement for true reflectivity determination is the assumption that the light emitted by the flash lamps propagates as

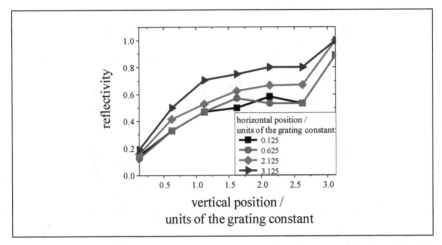

Figure 58: As-measured reflectivity for increasing vertical distance of the lamp detector
w.r.t. the grid of the mechanical oscillator. The values in the legend describe
the distance of the wafer detector from the mechanical oscillator in units of
the grating constant.

a plane wave. This is not generally true as the lamps can be approximated to
be point sources in close proximity. In fact, the distance between source
(lamps) and wafer is kept small in order to prevent radiation losses. To gain
constant plane wave illumination of the wafer the chamber walls of the
annealer may be diffuse reflective. However, this will lead to intensity loss,
though it will improve uniform tempering of the wafer. In fact, the chamber
walls of a typical annealer are highly reflective to prevent intensity loss.
Though this supports the point source nature described above in contrast to
diffusive-type chamber walls, it effectively enlarges the lamp bank and thus
increases the as-measured reflectivity towards its real value.

From $I_{thermal}$ the temperature is conventionally inferred by using Planck's
law for black bodies (cf. Equation (19) and the sample emissivity, which defines the ratio between the thermal radiation of a real and a black body. As
Kirchhoff's law might not be readily applied for millisecond annealing for
reasons discussed in section 3.2, emissivity deserves special attention. The
sensitivity on temperature for changing emissivity and false reflectivity
measurement can be found in Figure 23 and in Figure 59. The detection
wavelength regime provides a much stronger contribution to the exact
temperature determination (cf. Figure 59) though a change in detection

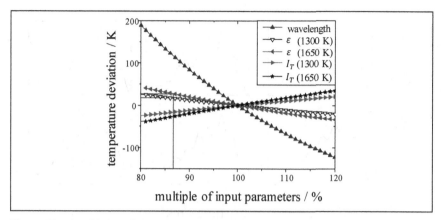

Figure 59: Sensitivity of the temperature deviation on $I_{thermal}$ (here: I_T), ε and λ_{det} (here: wavelength). The calculation is based on Planck's law (cf. Equation (19)) at a central wavelength of 950 nm.

wavelength is not obvious at first hand when neglecting ageing of the detector. However, the detector responsivity, which is a function of wavelength, may not change.

Figure 60 shows the online measurements of the modulated flash pulse after reflection off the wafer surface with a Silicon photodiode (wavelength detection range between 300 nm … 1100 nm). The respective amplitude

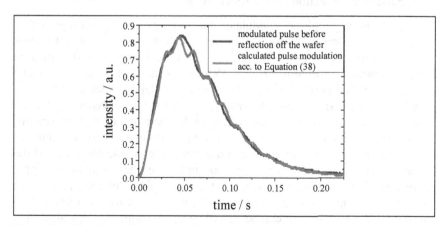

Figure 60: Modulated flash pulse with a time duration of 80 ms at FWHM after reflection off the wafer surface (detector: silicon photodiode with a wavelength regime between 300 nm and 1100 nm).

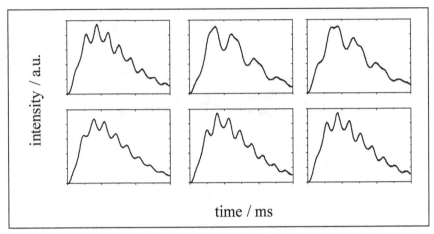

Figure 61: Reproducibility of ripple formation on flash pulses with a time duration of 80 ms at FWHM. Total recorded time comprises 150 ms. Divisions on the ordinate are identical.

modulation on the incident flash pulse can be found in Figure 51. The measurement is compared to the calculated amplitude-modulated flash pulse according to Equations (38) and (39). In Figure 51 the modulation of the flash before reflection could well be matched by Equation (38). However, stray radiation leads to a poorer match for the reproduction of the modulated flash pulse after reflection off the wafer surface.

Figure 61 shows amplitude-modulated flash energy density profiles with a pulse time duration of 80 ms at constant incident energy density to show reproducibility of the ripple formation. The modulation is shifted in time w.r.t. the mutual position of grid and detector at the start of the flash pulse. At first sight, one could find that the modulation depth varies despite equal conditions for all shown pulses. However, a closer look reveals that for the two profiles at the top right corner of Figure 61 there are actually two modulations depths. In other words, every other ripple is equally deep and one can find one fixed value for the modulation depth at the double frequency of the other profiles. However, this does not influence the strength of ripple pyrometry. Contrariwise, it shows that the actual shape of the ripple is not important as long as it is ensured that the reflection of the modulated flash pulse off the wafer surface can be described by the same ripple. Calculation of the reflectivities was performed using Equation (39) and for reference,

Table 4:   Determination of the reflectivity for weakly doped p-type Silicon wafers (dopant concentration $< 10^{15}$ cm^{-3}), shallow Boron-doped Silicon Si:B (implantation dose $< 10^{21}$ cm^{-2}, implantation energy: 500 eV) and SiO$_2$ on a Silicon substrate at RT.

Wafer material	Reflectivity obtained by ripple pyrometry	Reflectivity obtained by reflectometry measurements
Silicon	0.391 ± 0.052	0.372 ± 0.001
Si:B	0.404 ± 0.051	0.372 ± 0.001
SiO$_2$	0.252 ± 0.047	0.203 ± 0.001

reflectometry measurements were also carried out for the detection wavelength regime. D_L and D_W were determined by comparison of the lamp and the wafer detector for equal incident flash energy density. The result was verified online by the fact that the ratio between D_L and D_W equals the ratio between I_L and I_W. ΔI_L and ΔI_W were determined from the fit according to Equation (38).

Table 4 summarises the obtained reflectivities from ripple pyrometry and reflectometry for weakly doped p-type Silicon (dopant concentration $< 10^{15}$ cm^{-3} will be referred to as *Silicon* only in the following), shallow boron doped silicon (implantation dose of 10^{15} cm^{-2} at an implantation energy of 500 eV) and SiO$_2$ on a Silicon substrate. The values obtained from reflectometry measurements are smaller than those from ripple pyrometry, yet they range within the standard deviation of the results obtained by ripple pyrometry. Surprisingly, the shallow implantation does not have a significant contribution towards its reflectivity despite considerably high dopant concentrations already at medium doses as expected from the findings in subsection 2.3.3.

The uncertainty in reflectivity determination σ_R by ripple pyrometry is found to be 13.3 % for Silicon, 12.6 % for shallow Boron-doped Silicon (Si:B) and 18.7 % for SiO$_2$ on Silicon. These values agree with the prediction made by Figure 59. σ_R decreases rapidly in close vicinity to the wafer, but only reaches the real value of the sample reflectivity if both, the lamp and the wafer detector, are placed on the wafer surface which cannot be achieved in reality with the same accuracy due to the final size of the detectors.

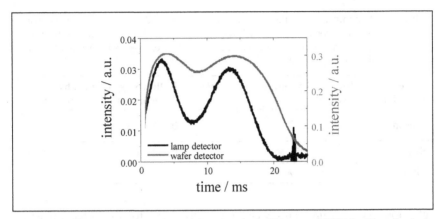

Figure 62: Flash pulse profile with a time duration of 20 ms at FWHM with an intrinsic
ripple, which was created by a suitable coil-capacitor-arrangement (cf.
Figure 35).

As indicated before, an intrinsic ripple due to a suitable coil-capacitor-
arrangement can also be used to determine the sample reflectivity. For this
purpose, a modulated 20 ms FWHM flash pulse was monitored before and
after reflection off the wafer surface. The pulse profile can be found in
Figure 62. The ripple size of the wafer detector is decreased compared to the
one of the lamp detector as expected. The ratio of the relative – to compensate for the lack of knowledge in detector sensitivity – ripple sizes $\frac{\Delta I_W}{I_W}$
and $\frac{\Delta I_L}{I_L}$ was investigated to obtain a measure of the sample reflectivity. The
flash pulse was monitored by a Silicon photodiode with a wavelength
detection range between 300 nm and 1100 nm. The sample reflectivity was
determined for incident flash energy densities, which do not lead to
significant temperature rise and was found to be 0.171 ± 0.025. This
compares to the values given in Table 4 to be just above 50 % of the
expected value. This may occur due to a superposition between the reflected
flash radiation off the wafer surface and direct flash radiation being multiple
times reflected off the chamber walls. Yet, compared to the results from
Table 4 reflectivity determination by intrinsic means of amplitude modulation are less suitable than mechanical ripple pyrometry.

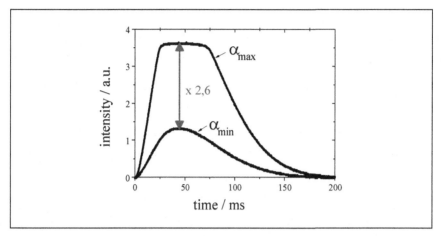

Figure 63: Flash lamp irradiation for maximum (< 90°) and minimum angles (> 0°) w.r.t. the wafer normal due to the thermal shadow effect (Figure 47). At angles of 0 ° and 90 ° no flash light intensity reaches the active area of the detector.

The thermal radiation of a hot Silicon wafer does not depend strongly on the angle due to the high refractive index of Silicon [**54**]. However, because of the sole presence of the detector, the angle of detection needs to be fixed for repeatability. The thermal shadow caused by the physical mass of the detector which acts a heat sink will decrease with increasing angle by $l_{det} \times cos\alpha$ following simple geometrical considerations (cf. Figure 47; note: Figure 47 studies the influence of the angle of the lamp detector, while the angle of the wafer detector is considered here) [**77**]. However, the field of view w.r.t. stray radiation will increase with increasing angle, thereby increasing the as-measured incident radiation on the detector as shown in Figure 63. Thus, small angles are generally favourable although error insertion due to the thermal mass of the detector needs to be reduced. Detailed information can be found in [**59**].

6 Experiments – Ripple Pyrometry during Flash Lamp Annealing

6.1 Proof of Concept

This chapter intends to investigate the suitability of ripple pyrometry for reliable temperature measurement. Clearly, without background correction the power density of the as-measured wafer radiation will lead to a much higher temperature reading due to the superposition of reflected flash lamp radiation (background) and thermal radiation. However, this variation is not determined by a fixed value, which can be subtracted from the as-measured power density, but this value varies due to the non-linear increase in power density in order to achieve rising wafer temperatures (cf. Table 1). Figure 64 shows the expected detection signal in contrast to the background-corrected thermal radiation.

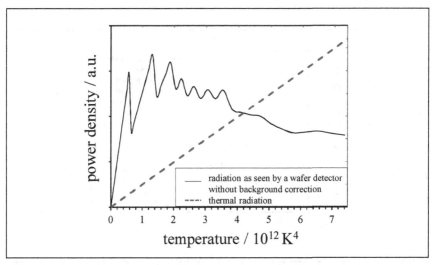

Figure 64: Demonstration of the superiority of background-corrected measurements of thermal radiation (dotted line) versus non-corrected measurements (solid line). The data is created acc. to Table 1. The solid graph is displayed one order of magnitude smaller for better demonstration.

The solid graph is made of a random set of data, which includes thermal radiation and exactly this amount of flash lamp radiation acc. to Table 1, which is necessary to create this thermal radiation (dotted line). In other words, if the sample reflectivity and thus the amount of flash lamp radiation that is reflected off the wafer surface, is unknown, the solid graph is measured leading to false temperature readings. Figure 64 proves that online background-correction needs to be addressed if considering temperature measurement at all. In fact, the non-corrected data is reduced by one order of magnitude in power density to show the non-linearity. The extraordinary large background during FLA would falsify the temperature measurement even worse (cf. Table 1).

The knowledge of the reflectivity of the sample gives a measure for the background. However, it also requires the ability to measure the reflectivity to such a precision that the weak thermal radiation can be calculated within a few K, which is complicated by the fact that thermal radiation is much smaller than the background radiation from the flash lamps as indicated in Table 1. Figure 65 shows the power density for various temperatures and

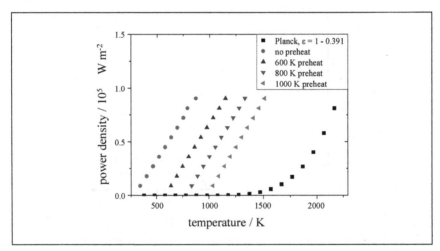

Figure 65:   Thermal radiation of Silicon according to Planck's law (cf. Equation (19)) in a wavelength regime between 800 nm and 1100 nm (squares); and uncertainty in power density of the reflected flash pulse due to the uncertainty in sample reflectivity in Table 4. The power density of the incoming flash lamp radiation was determined by numerical simulations acc. to subsection 5.2.1 such that to give the respective temperature reading.

compares them with the power density arising from the uncertainty in sample reflectivity as summarized in Table 4. For the calculation the necessary power density to heat the wafer in 20 ms up to the desired temperature was multiplied with the uncertainty in sample reflectivity from Table 4 for Silicon and plotted in Figure 65 for various preheat temperatures. The figure shows that the resulting uncertainty in power density is larger than the power density due to thermal radiation, which was obtained from applying Planck's law. This means, the resulting uncertainties of temperature measurement are larger the less precise the determination of the sample reflectivity. As shown in Table 1, a precision of 0.1 % is sufficient for temperatures above 1300 K. However, a much larger uncertainty of online reflectivity measurement of about 13.3 % for Silicon was obtained (cf. Table 4). According to Figure 59, the resulting temperature uncertainty amounts to at least 15 K at 1300 K and 23 K at 1650 K (note: the dotted lines in Figure 59 serve as guidance to the eye). Therefore, a lower limit theoretical temperature uncertainty of 15 K can be given for the temperature regime of interest.

Additionally, during preheating a steady gas flow to ensure an inert and well-defined atmosphere can cause large uncertainties in T. The calculation shown in Figure 66 was made assuming the wafer to be at a constant intermediate temperature T as indicated. The temperature decrease

$$\frac{\Delta T}{\Delta t} = \frac{\alpha A \, (T-300)}{m_{Si} c_{Si}}, \text{ where}$$

$$\alpha = 5.6 + 4 \sqrt{\frac{2p}{\rho_{gas}}} \tag{42}$$

of the single-crystal Silicon wafer (mass m_{Si}, heat capacity c_{Si}) in K/s is shown for a constant gas flow of Nitrogen and Argon with density ρ_{gas} at a pressure p over the entire wafer surface A. The calculation assumes that the gas, which is at 300 K, does not heat up. In fact, during the short-time span of FLA, a significant contribution of the gas flow to the temperature measurement uncertainty may not be expected. Yet, the incident flash energy density only determines the possible rise in wafer surface temperature (cf. Equation (4)). Therefore the peak temperature during FLA depends on the preheat temperature and shows that without an online technique for temperature measurement, the gas flow may lead to erroneous processing. A decrease of flash energy density by absorption of the gas, however, can be neglected due to the small absorptivity in the spectrum of the flash radiation.

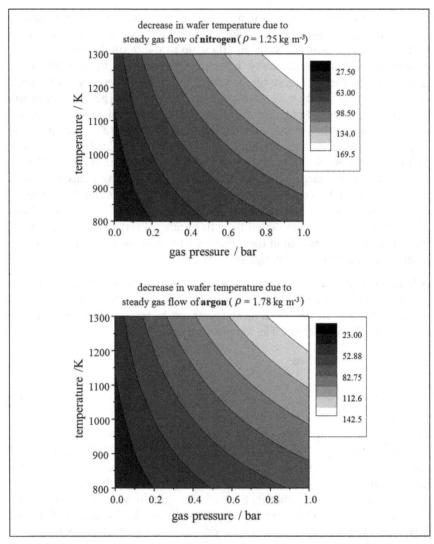

Figure 66: Temperature decrease in K/s due to a steady gas flow of Nitrogen (top) and Argon (bottom) as a function of gas pressure and preheat temperature.

6.2   Low Temperature Measurement by Ripple Pyrometry

The concept of ripple pyrometry for temperature measurement was first tested at low temperatures < 1000 K. For this regime, Aluminium and Gold dots were deposited on a Silicon wafer as shown in Figure 67. The dots were 50 nm in thickness and 500 µm in lateral size. It is important to note that with these values the dots do not exceed a critical thickness such that the metal can be melted within the short period of time during millisecond annealing. The idea was to place the dots in such a way that the overall area of the metal dots is negligible in comparison to the wafer surface. This means that the optical parameters of the wafer substrate, namely Silicon, are applicable instead of the metal parameters. For this assumption to hold during the measurement the pilot laser of the pyrometer was exploited to ensure that the detection area of the pyrometer was placed between the dots.

The Aluminium dots were sputtered and the Gold dots were fabricated onto the Silicon by evaporation deposition. Prior to this, the wafers were cleaned with H_2O_2 / H_2SO_4 to remove the thermally insulating native oxide layer on the Silicon surface immediately prior to the coating to ensure that the dots are not melted by direct flash lamp irradiation, but rather by indirect heating from the Silicon substrate. The metal and the Silicon form an eutectic, which lowers the melting point. Table 5 lists the melting points of each eutectic and compares the values with the individual melting point of the metals. It also gives the fraction of the metal in the Silicon-metal-eutectic as well as the reflectivities of pure Gold and Aluminium [78], [79], [80], [81].

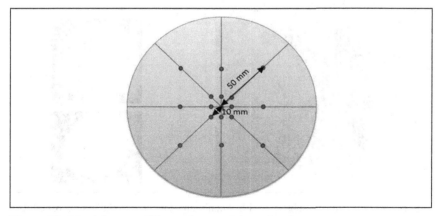

Figure 67:   Metal dots on a Silicon substrate to determine the temperature.

Table 5:   Temperature measurement for the low-temperature regime: overview on metal parameters. The reflectivity is given for a wavelength regime between 800 nm and 1100 nm.

	Fraction of metal in eutectic	Eutectic melting point	Melting point of the pure metal	Reflectivity of the metal
Aluminium	12.2 at.%	(850 ± 1) K [78]	933 K	0.913 [79]
Gold	18.6 at.%	(636 ± 3) K [80]	1337 K	0.977 [81]

The melting of the metal dots was observed by light microscopy to follow the melting process. The dots were said to be melted if their shape changed significantly (cf. Figure 68). The annealing took place without preheating for a flash pulse with a time duration of 20 ms. To calculate the amount of thermal radiation according to Equation (40) I_W was measured by the pyrometer with a filter of an optical density of 3.0 (attenuation factor $= 10^{-3.0} = 1000$), which was presented in section 5.3. I_L had been determined before the annealing process by an energy density meter by the laser components company *Laser2000™*. According to Equation (39) the temperature of the silicon can be inferred from the modulation depth ΔI on the lamp and wafer detector signal I_L and I_W knowing the respective detector responsivities D_L and D_W. D_L and D_W were obtained as described in section 5.6. Yet, due to technical limitations, an online determination of the

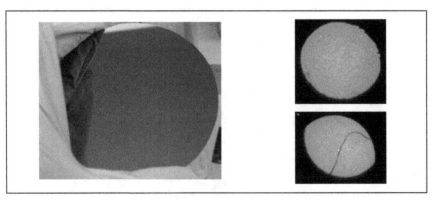

Figure 68:   Left: Silicon substrate with Aluminium dots. Top right: As-deposited Aluminium dot Bottom right: Melted Aluminium dot

reflectivity was not possible. This, however, does not affect the concept of ripple pyrometry, which only refers to the fact that an amplitude modulation of the lamp radiation is generated to distinguish thermal radiation from the latter. Whether this is done off- or online is not decisive. Further, literature shows that an offline determined reflectivity by a Silicon detector closely meets the online control data [82], which is further supported by Figure 18. Despite the increase of the reflectivity for wavelengths around the Silicon bandgap at 1100 nm, the dependence on temperature is not as pronounced for the given wavelength detection regime of the pyrometer above RT. However, there is a significant change in reflectivity at RT (cf. Figure 18). Yet, the resulting difference in reflectivity w.r.t. elevated temperatures is less than the uncertainty obtained from reflectivity determination by ripple pyrometry (cf. Table 4). Therefore, the afore determined reflectivity by reflectometry (cf. Table 4) was used to calculate the amount of flash radiation in the wafer detector signal I_W. To include the detection regime of the pyrometer, this value has been corrected for a wavelength range of 800 nm ... 1100 nm to yield a reflectivity of 0.324 ± 0.001.

Further, the emissivity of the Silicon needs to be determined. For this purpose, the question arises whether Kirchoff's law may be applied. Generally, it is not valid outside thermal equilibrium (cf. section 3.2). According to numerical simulations the slope of the thermal profile is 13 K/ms (cf. Figure 69), the exposure time of the pyrometer, however, is 2 ms. Therefore, there is a jump in temperature of 26 K for each sampling, which represents 8 % of the total temperature increase during annealing, which cannot be considered as negligible anymore and needs to be taken into account for further discussions.

Using Planck's law between wavelengths of 800 nm and 1100 nm, the thermal profile of I_W was determined and is displayed in Figure 69 along with numerical simulations ("calculated temperature") according to sub-section 5.2.1. The starting values, however, cannot be compared due to the lower limit temperature response of the pyrometer of 873 K. The temperatures obtained from the Silicon wafers with Gold (dotted graph) and those with Aluminium dots (solid graph) are comparable in dimension as expected from the calculated values. The small signal-to-background ratio is the reason for the as-measured detector response to follow the course of the incident flash pulse rather than the true temperature evolution as determined by simulation. The contribution from thermal radiation may only be visible

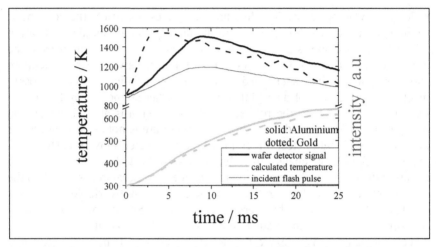

Figure 69: Comparison of the wafer detector signal and the simulated surface temperature for a flash pulse with a time duration of 20 ms at FWHM and with an incident energy density of 3.518 ± 0.194 J/cm² (Aluminium) and
3.240 ± 0.198 J/cm² (Gold). The course of the incident flash pulse is shown
for comparison.

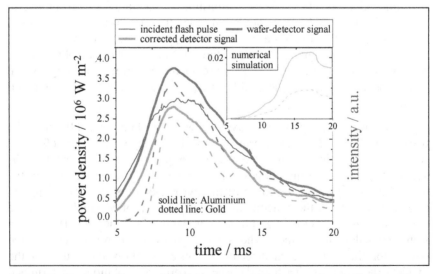

Figure 70: Background correction of the as-measured detector signal. The inset shows
the course of the power density according to Planck (cf. Equation (19)) obtained from numerical simulations of the surface temperature.

in the falling edge of the flash pulse. This is possible due to the time shift between the temperature rise and the rising edge of the incident flash pulse.

According to Equation (40) the thermal radiation $I_{thermal}$ was subsequently obtained and the result of the background correction is shown in Figure 70. The inset shows the power density according to Planck's law of the simulated thermal profile. The background-corrected graph is still many orders of magnitude larger than the power density of the thermal radiation (see inset).

The as-measured signals were translated into spectral energy density for the above given wavelength regime using Planck's law and then integrated to obtain the total energy density of the wafer detector (= I_W = 3rd column in Table 6). The results are compared to the incident flash energy density (= I_L = 1st column in Table 6) and show that the ratio of as-measured energy density I_W towards the incident energy density I_L is much larger than expected from the reflectivity measurements in Table 4 (see 2nd column in Table 6). As discussed before, a change in reflectivity is indeed expected when moving from RT to higher temperatures. However, the reflectivity should decrease (cf. Figure 18).

Of course, this comparison can only give a quick overview on the issue as the ratio between I_W and I_L does not yield the sample reflectivity as the thermal radiation needs to be taken into account. However, in this particular case the thermal radiation is almost negligible w.r.t. I_W. Keeping in mind that the spectral energy density of the (simulated) thermal radiation is too small to balance the difference between the 2nd and the 4th column in Table 6, the reason for this result needs to be found elsewhere. Stray radiation from the flash lamps ought to be taken into account although the pyrometer should be protected from direct lamp radiation at its position above the chamber (cf. Figure 42; note: Figure 60 was not measured by the pyrometer, but by the Silicon photodiode, which was placed inside (!) the chamber). Multiple reflections from the side walls of the chamber should not play a role as the pyrometer was placed outside the chamber in such a way that it viewed the wafer through a pinhole. However, as for the intrinsic ripple studied in section 5.6, they would explain the large as-measured reflectivity. The last column in Table 6 indicates the total energy density of the simulated tempe-rature distribution. The temperature simulation assumes no direct absorption of the metal dots which is reasonable w.r.t. the reflectivity values given in

Table 6:   Overview of the energy densities of incident, reflected and thermal radiation for the low-temperature regime. All values were adapted to the wavelength detection regime of the pyrometer (800 ... 1100 nm). The thermal energy density created during the flash anneal is obtained from numerical simulations according to subsection 5.2.1.

	Aluminium	**Gold**
Flash energy density in J/cm² (lamp detector)	3.518 ± 0.194	3.240 ± 0.198
Reflected flash energy density in J/cm² (cf. Table 4)	1.140 ± 0.063	1.050 ± 0.064
As-measured energy density in J/cm² (wafer detector)	1.455 ± 0.063	1.632 ± 0.203
As-measured reflectivity	0.414 ± 0.011	0.504 ± 0.010
Thermal energy density in J/cm²	4.81×10^{-8}	1.33×10^{-8}

Table 5. Assuming direct absorption of the metal dots having a higher reflectivity than Silicon, the incident flash energy would not have been sufficient for dot melting. The thermal energy density is many orders of magnitude smaller than the uncertainty in determining the energy density of the reflected flash. It seems that the pyrometer still sees direct flash light despite extensive precautions.

Comparing the reflected energy density from Table 6 (cf. 2nd column) to the as-measured wafer detector signal I_W (cf. 3rd column) gives the amount of thermal radiation as determined by this method assuming no other factors contribute to the larger radiation being viewed by the wafer detector w.r.t. the expected reflected flash energy density. Neglecting the thermal profile in favour of constant annealing temperature, which is reasonable with hindsight on the measurement results, a temperature reading for the 20 ms flash pulse was obtained using Planck's law (cf. Equation (19)), which yields for the measurement using Aluminium dots (2698 ± 15) K and for the Gold dots (2879 ± 140) K. These large temperatures can be explained by the low signal-to-background ratio at lower temperatures (cf. Figure 65) in combination with the considerations on stray radiation and multiple reflections. As reflectometry results are employed the uncertainty in the determination of R (cf. Table 4) does not contribute.

These results suggest that measurement at elevated temperatures needs to be taken into consideration as discussed before.

6.3 High Temperature Measurement by Ripple Pyrometry

6.3.1   Ripple Pyrometry on Shallow Boron-Doped Silicon

For the purpose of high temperature studies > 1000 K, measurements were performed on shallow Boron-doped Silicon samples. The samples were implanted at an energy of 500 eV and at a dose of 10^{15} cm^{-2}. Figure 71 shows simulation results for the depth distribution of the Boron dopant in Silicon [28]. The results need to be multiplied by the implantation dose to yield the dopant depth concentration after implantation. From the maximum number of atoms per cm^3 in Figure 71: per given implantation dose, a maximum dopant concentration in the order of 10^{21} cm^{-3} is obtained taking into account the implantation dose given above. The mean projected range is in the order of 40 Å [28].

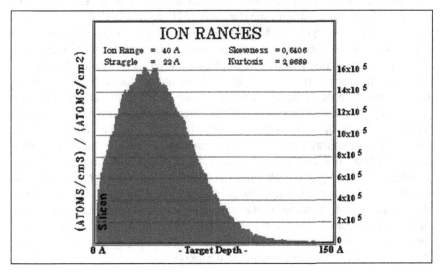

Figure 71:   Ion ranges for Boron-doping into Silicon at an implantation energy of 500 eV (figure taken from [28]).

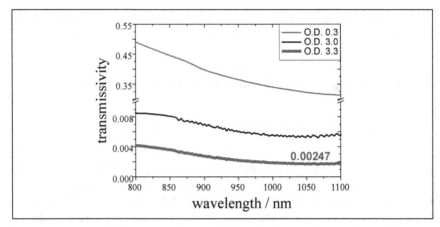

Figure 72: Transmission spectra of the single neutral density filters used and their superposition (O.D. 3.3); O.D. = optical density. The number notes the integral for
the wavelength regime of the detector between wavelengths of 800 nm
and 1100 nm.

The samples were preheated with halogen lamps up to a temperature of
773 K. A flash pulse with a time duration of 20 ms at FWHM was used at a
nominal energy density of 87.6 J/cm². Five samples were studied under
similar process conditions as described above. The samples were viewed by
the pyrometer through two neutral density filters of optical density 0.3
(attenuation factor = $10^{-0.3}$ = 2) and 3.0 (attenuation factor = $10^{-3.0}$ = 1000),
respectively, to protect the sensitive pyrometer optics from the high-intensity
flash. The transmission spectrum of the filters used is shown in Figure 72.
Integration over the detection spectrum yields an attenuation of 0.247 %.
Temperature measurement was again undergone on the basis of Equations (39) and (40). Just as before for the low temperature regime (cf. section 6.2), D_L and D_W were obtained according to the procedure explained in
section 5.6. The technical limitations that excluded online determination of
the reflectivity of the wafer surface also apply here. Therefore, the reflectometry results from Table 4 were used again corrected by the detection wavelength regime of the pyrometer (800 nm … 1100 nm) to give a reflectivity of
0.324 ± 0.001 for shallow Boron-doped Silicon.

Figure 73 compares the thermal profile obtained from numerical simulations
according to subsection 5.2.1 and the as-measured thermal profile I_W. The
steep rise of the measured flash pulse is probably just an artefact of the lower

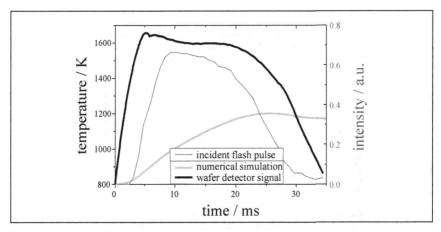

Figure 73: Comparison of the wafer detector signal and the simulated surface temperature for a flash pulse with a time duration of 20 ms at FWHM and an incident energy density of 4.575 ± 0.282 J/cm².

limit sensitivity of the pyrometer < 873 K. This assumption is supported by comparison to the profiles in section 6.2. The broadening of the wafer detector signal I_W still cannot be explained by thermal radiation due to its low share in I_W, but by geometrical considerations of the annealing chamber.

Figure 74 shows the result from background-correction and the inset shows the power density of the simulated thermal profile. The absolute values are compared in Table 7. Clearly, again the corrected signal cannot verify the calculated thermal profile. Similar to the results from the low temperature regime in section 6.2. The energy density of the as-measured radiation on the wafer detector I_W is much higher than expected from the sample reflectivity and the temperature simulations. This supports the assumption from section 6.2 that stray radiation from the flash lamps still contributes to the detector signal despite suitable placing of the pyrometer outside the chamber. Multiple reflections also need to be considered here.

Interestingly, the as-measured wafer detector signal I_W exceeds the incident radiation signal I_L. Therefore, the as-measured reflectivity is found to be larger *one*. The origin of this effect is not fully understood, yet. The measurement of the incident flash energy density, however, is reliable as it reproduces the expected temperature readings (cf. subsection 6.3.2). Also, stray radiation may not increase I_W beyond I_L. However, multiple reflections may lead to increased pyrometer readings.

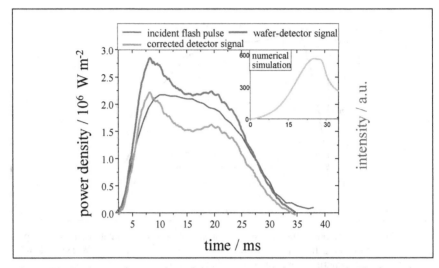

Figure 74: Background correction of the as-measured detector signal. The inset shows the course of the power density according to Planck (cf. Equation (19)) obtained from numerical simulations of the surface temperature.

Similarly to Table 6 a temperature reading can be obtained from the wafer detector response I_W and the calculated value for the reflected flash radiation (cf. 2nd column) resulting in (2989 ± 268) K for a 20 ms flash pulse assuming constant process temperature. The results are similar to the ones derived in section 6.2, which emphasises the negligibility of the thermal radiation w.r.t. the background of the flash lamps.

Table 7: Overview of the energy densities of incident, reflected and thermal radiation for the high-temperature regime. The thermal energy density is obtained from numerical simulations. All values were adapted to the wavelength detection regime of the pyrometer (800 … 1100 nm).

Flash energy density in J/cm² (lamp detector)	4.575 ± 0.282
Reflected flash energy density in J/cm² (cf. Table 4)	1.482 ± 0.091
As-measured energy density in J/cm² (wafer detector)	5.343 ± 1.025
As-measured reflectivity	1.168 ± 0.208
Thermal energy density in J/cm²	$1.13 \times 10\text{-}3$

6.3.2   Temperature Control – Secondary Ion Mass Spectrometry

In addition to numerical temperature simulations, which cannot reflect the full experimental reality, the concentration profile of the diffused dopants was measured afterwards by SIMS and compared to numerical SIMS simulations for the verification of the measured thermal profile during annealing. Since dopant diffusion is a function of temperature (cf. Equation (18b)) the thus obtained concentration profile reflects information on the thermal profile as well.

A rough first calculation of the diffusion of Boron into Silicon assuming a box-shaped thermal profile during annealing with a width of 30 ms, which compares to a profile with a duration of 20 ms at FWHM, can be found in Figure 75. The simulations were carried out at the *IISB Fraunhofer Institute* (Erlangen, Germany). The underlying model considers the evolution of self-interstitial clusters as well as BICs of various stoichiometric combinations. Details are given elsewhere [*40*].

The simulated results were compared to SIMS measurements (cf. Figure 76) at *IHP Microelectronics* (Frankfurt/Oder, Germany) using an oxygen source. For reproducibility studies, five identical samples were annealed and investigated. They are referred to with an "S" in Figure 76; an "R" means that SIMS was performed close to the wafer edge. Results from FLA coupled to conventional RTA at a temperature of 1573 K are labelled "fRTA_1300". In this case, RTA is used as pre-heating and FLA for annealing. The as-implanted sample is referred to by "BOP5_1E15ai". The abbreviation "3OSi" indicates the standard for the measurement.

Comparing the simulations in Figure 75 to the measurement in Figure 76 agrees with the numerical simulations of the process temperature (cf. Figure 73) and falsifies the obtained temperature readings from the measurement as expected. As predicted, the wafer surface just reached the desired temperature for significant diffusion of the Boron into the bulk during annealing with the parameters as specified in the preceding subchapter. The results from "fRTA" show less diffusion into the bulk than for FLA ("S"), which may be attributed to the longer and compared to fRTA uncontrolled pre-heating in this case.

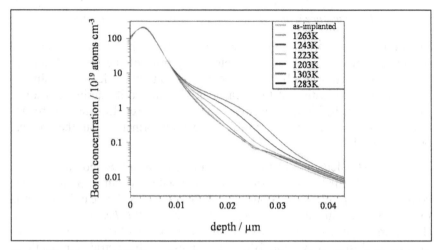

Figure 75: Simulated SIMS profiles for a box shaped thermal profile and for the indicated temperatures and an annealing time of 30 ms. Simulations were performed by Hackenberg, M. [*83*].

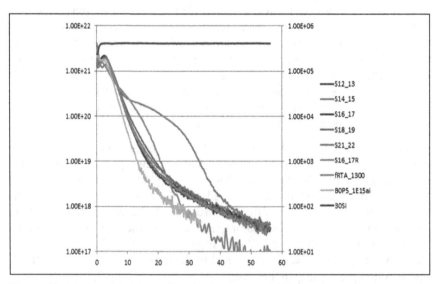

Figure 76: Measured SIMS profiles for Boron-doped Silicon after FLA at an energy density of 4.575 ± 0.282 J/cm² and for flash pulse times of 20 ms at FWHM; S = sample, BOP5_1E15ai = as-implanted wafer, 3OSi = internal standard, fRTA_1300 = combination of FLA and RTA at a peak temperature of 1573 K. Measurements were performed by Bolze, D. [84].

The results from chapter 6 confirm, that for increasing temperature the ratio of thermal radiation and reflected flash lamp radiation grows in favour of the first (cf. Figure 65). Although stray radiation seems to be a major problem, the uncertainty in determining the temperature is expected to decrease for elevated temperatures. However, temperature measurement could not be repeated for elevated temperatures above 1200 K due to technical limitations. Another approach to counteract stray radiation is to enlarge the mechanical oscillator.

7 Closing Discussion and Outlook

Summarizing, FLA is a powerful and efficient tool for thermal treatment of a variety of materials, ranging from microelectronic to photovoltaic applications. It holds a golden balance between laser and furnace annealing with respect to dopant activation while minimising dopant diffusion. Moreover, being a large area covering annealing tool, FLA saves costs and time. Compared to halogen lamps, which emit light with photon energies mostly below the Silicon bandgap, the UV/Vis spectrum of the flash lamps (cf. Figure 22) also shows the superiority of FLA over conventional thermal treatment.

To the best of the authors knowledge there is no alternative to filterless background-corrected temperature measurement during FLA. However, due to the low signal-to-background ratio, where *signal* refers to the thermal radiation of the hot wafer surface, background correction is essential for trustable temperature readings. Not only the present thesis has demonstrated the necessity of ripple pyrometry to FLA (cf. chapter 4), but also it has shown its suitability for flash pulse times as short as 1 ms (cf. section 5.5), especially with regard to the possibility of engineering the modulation frequency of the detected signal by the variation of the grid displacement s w.r.t. the grating constant g. The modulation has been shown to be predictable by knowing the system parameters of the mechanical oscillator, which makes the present technique reproducible and controllable.

Conventional pyrometry has not been able to obtain true temperature readings due to the high background which is created by the flash lamps. Therefore, industrially relevant online closed-loop process control to act on changes during annealing could not be performed. Ripple pyrometry, however, has been shown to provide an online measure for the degree of background radiation which had not been accessible before.

The conventional way of performing amplitude modulation for ripple pyrometry, namely through making use of the net frequency when operating halogen lamps, is not applicable to FLA. Thus the question arose whether ripple pyrometry is suitable for FLA at all. An answer was found in the loudspeaker technology, i.e. an AC driven "voice" coil is moved under the influence of a magnetic field due to the Lorentz force. In the case of a loudspeaker, this coil moves a thin membrane; for motor actuation the coil drives a mass, here: a specially prepared grooved quartz plate (cf. section

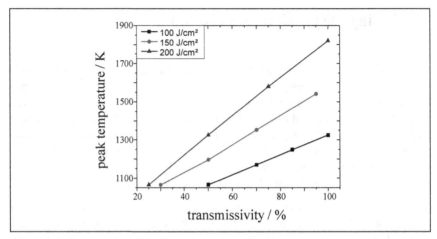

Figure 77: Deviation in maximum temperature due to varying grid transmissivity τ as a function of incident flash energy density for flash pulse times of 20 ms at FWHM. The calculation is based on Planck's law (Equation (19)).

5.3). This grid leads to an amplitude modulation of the flash and thus is used for masking. The total loss of irradiance due to the grid can be approximated by τA_τ, where A_τ denotes the percental area covered by the grid and τ the transmissivity of the grid. τA_τ leads to a reduction in maximum temperature according to Figure 77. However due to the constant horizontal movement of the grid no local thermal shadow is to be expected, which is of much more importance for industrial applications.

The parameter which aggravates the application of this technology for short flash pulses, i.e. for higher frequencies up to several 1000 Hz, is mainly the mass of the quartz plate. However, this work could show that to the detector the proposed method for amplitude modulation of the incident flash pulse can generate a multiplication of the input frequency (cf. Figure 54). Moreover, it was found that the amplitude modulation of the flash lamps does not influence significantly the thermal radiation of the wafer (cf. section 5.5).

For temperature measurement, the ability of ripple pyrometry for online reflectivity measurement during FLA has been studied. The lower limit uncertainty in reflectivity measurements has been found for Silicon with 15 K temperature error at 1300 K and is thus eight times smaller than offline techniques. The results of raytracing simulations show that this uncertainty agrees with theory. The reflectivity can be best determined in close proxymity to the wafer such that the light path for both detectors has approximately the same length. This thesis has shown that the here presented method can give the true sample reflectivity (cf. section 5.6).

By online determination of the sample reflectivity, any changes in the optical properties of the material, such as phase changes, can be observed. In particular, melting of the Silicon surface and recrystallization during annealing can be accounted for during online process control. Therefore, it is reasonable to call ripple pyrometry an online material-independent method.

The influences of detector position, size and angle as well as chamber and oscillator geometry on the determination of the wafer reflectivity have been studied (cf. section 5.4 and 5.6).

Two sets of experiments were put up to proof the concept of ripple pyrometry for the low- and high-temperature regime. SIMS measurements and simulations were taken into consideration for reference. Both results confirmed the findings from theoretical considerations that this technique is most powerful for temperatures > 1300 K with a higher signal-to-background ratio of thermal radiation towards lamp radiation (background). Multiple reflections inside the chamber and stray radiation also ought to be considered. An enlargement of the mechanical oscillator beyond the size of the lamp bank may improve the results.

Bibliography

1   Atkins, P. W. and de Paula, Julio. Physikalische Chemie. Oxford University Press, Oxford, 2013.

2   Sedgwick, T. O. Short Time Annealing. Journal of the Electrochemical Society, 130, 2 (1983), 484-493.

3   Reichel, D. et al. Temperature Measurement in Rapid Thermal Processing with focus on the application to Flash Lamp Annealing. Critical Reviews in Solid State and Material Sciences, 36, 2 (2011), 102-128.

4   Stuart, G. C. et al. Temperature diagostics for a dual-arc FRTP tool. (Vancouver, Canada 2002), International Conference on Advanced Thermal Processing of Semiconductors.

5   Walk, H and Theiler, T. The wafer temperature measurement in dual OH-Band quartz tube. (Monterey, California 1994), International Conference on Advanced Thermal Processing of Semiconductors.

6   Schietinger, C.W. and Adams, B. E. Non contact technique for measuring temperature of radiation-heated objects. USA Patent US 5318362 (1994)

7   Cezairlizan, A. et al. Issues and future directions in subsecond thermophysics research. International Journal of Thermophysics, 11, 4 (1990), 819-33.

8   Gebel, T. Flash Lamp Annealing with millisecond pulses for ultra-shallow boron profiles in silicon. Nuclear Instruments and Methods in Physics Research Section B: Beam Interactions with Materials and Atoms, 186, 1-4 (2002), 287-291.

9   Lanzerath, F. et al. Boron activation and diffusion in silicon and strained silicon-on-insulator by rapid thermal and flash lamp annealings. Journal of Applied Physics, 104, 4 (2008), 044908.

10 Kuzmany, H. Solid-state spectroscopy: an introduction. Springer-Verlag Berlin, Heidelberg, 2009.

11 Emmett, J. L. and Schawlow, A. L. Enhanced ultraviolet output from double-pulsed flash lamps. Applied Physics Letters, 2, 11 (1963), 204-205.

12 NSM Archive. Physical Properties of Semiconductors. IOFFE Physical Technical Institue. 2013.

13 Sun, B. K., Zhang, X., and Grigoropoulos, C. P. Spectral optical functions of silicon in the range of of 1.13-4.96 eV at elevated temperatures. International Journal of Heat an Mass Transfer, 40, 7 (1997), 1591-1600.

14 Green, M.A. and Keevers, M. Optical properties of intrinsic silicon at 300K. Progress in Photovoltaics, 3, 3 (1995), 189-192.

15   Agarwal, A. et al. Boron-enhanced diffusion of boron from ultralow-energy ion implantation. Applied Physics Letters, 74, 17 (1999), 2435-2437.

16   Schmid, P. E. Optical absorption in heavily doped silicon. Physical Review B, 23, 10 (1981), 5531-5536.

17   Basu, P. K. Theory of Optical Processes in Semiconductors: Bulk and Microstructures. Oxford Science Publications, Oxford, 1997.

18   Martienssen, W. Functional Materials - Semiconductors. In Springer Handbook of Condensed Matter and Materials Data. Springer Verlag, Berlin, 2005.

19   Okhotin, A. S. et al. Thermophysical Properties of Semiconductors. ATOM Publication House, Moscow, 1972.

20   Glassbrenner, C. J. and Slack, G.A. Thermal Conductivity of Silicon and Germanium from 3°K to the Melting Point. Physical Review, 134, 4A (1964), A1058-A1069.

21   Shanks, H. R. et al. Thermal Conductivity of Silicon from 300 to 1400°K. Physical Review, 130, 5 (1963), 1743-48.

22   Callister Jr., W. D. Materials Science And Engineering: An Introduction. John Wiley & Sons, Inc., Hoboken, 2003.

23   Fahey, P.M. et al. Point defects and dopant diffusion in silicon. Reviews of Modern Physics, 61, 2 (1989), 289-384.

24   Huitema, E. and van der Eerden, J. P. Defect formation during crystal growth. Journal of Crystal Growth, 166, 1-4 (1996), 141-145.

25   Voronkov, V. V. The mechanism of swirl defects formation in silicon. Journal of Crystal Growth, 59, 3 (1982), 625-643.

26   Xia, Y. Materials Chemistry. University of Washington, Department of Chemistry. 2005.

27   Newman, R. C. Defects in Silicon. Reports on Progress in Physics, 45 (1982), 1163-1210.

28   Biersack, J. P. Calculation of projected ranges — analytical solutions and a simple general algorithm. Nuclear Instruments and Methods, 182-283, 1 (1981), 199-206.

29   Agarwal, A. et al. Boron-enhanced-diffusion of boron: The limiting factor for ultrashallow junctions. In International Electron Devices Meeting (Washington, D.C. 1997), IEEE International Technical Digest.

30   Hong, S. N. et al. Characterization of Ultra-Shallow p+ -n Junction Diodes Fabricated by 500eV Boron-Ion Implantation. IEEE Transactions on Electron Devices, 38, 1 (1991), 28-31.

31   Schulz, M. Determination of Deep Trap Levels in Silicon Using Ion-Implantation and CV-Measurements. Applied Physics, 4, 3 (1974), 225-236.

32   Mehrer, H. Diffusion in Solids. Springer-Verlag Berlin Heidelberg, Heidelberg, 2007.

33   Robertson, L. S. Diffusion of ion implanted boron in silicon: the effects of lattice defects and co-implanted impurities. 2001.

34   Stolk, P.A. et al. Physical mechanisms of transient enhanced dopant diffusion in ion-implanted silicon. Journal of Applied Physics, 81, 9 (1997), 6031-6050.

35   Eaglesham, D. J. et al. Implantation and transient B diffusion in Si: The source of the interstitials. Applied Physics Letters, 65, 18 (1994), 2305-2307.

36   Griffin, P.B. and Plummer, J. D. Physical Modeling of transient enhanced diffusion in silicon. (Los Angeles 1996), Proceedings of the fourth international symposium on Process Physics and modeling in semiconductor technology, 101-115.

37   Michel, A. E. et al. Rapid annealing and the anomalous diffusion of ion implanted boron into silicon. Applied Physics Letters, 50, 7 (1987), 416-418.

38   Olesinsky, R. W. and Abbaschian, G. J. The B-Si (Boron-Silicon) System. Bulletin of Alloy Phase Diagrams, 5, 5 (1984), 478-484.

39   Duffy, R. et al. Boron uphill diffusion during ultrashallow junction formation. Applied Physics Letters, 82, 21 (2003), 3647-3649.

40   Pichler, P. et al. Process-Induced Diffusion Phenomena in Advanced CMOS Technologies. (2006), Trans Tech Publications.

41   Whelan, S. et al. Redistribution and electrical activation of ultralow energy implanted boron in silicon following laser annealing. Journal of Vacuum Science and Technology B, 20, 2 (2002), 644-649.

42   Larson, B. C. et al. Unidirectional contraction in boron-implanted laser-annealed silicon. Applied Physics Letters, 32, 12 (1978), 801-802.

43   Zhu, L. et al. Defect characterization in boron implanted silicon after flash lamp annealing. Nuclear Instruments and Methods in Physics Research B, 266 (2008), 2479-2482.

44   Longair, M.S. Theoretical Concepts in Physics. Cambridge University Press, Cambridge, 2003.

45   Kirchhoff, G. Untersuchungen über das Sonnenspektrum und die Spectren der chemischen Elemente. Abhandlungen der Königlichen Akademie der Wissenschaften zu Berlin, 227-240 (1862).

46   Kirchhoff, G. Über den Zusammenhang zwischen Emission und Absorption von Licht und Wärme. Monatsberichte der Königlichen Preußischen Akademie der Wissenschaften zu Berlin (1859), 783-787.

47   Boltzmann, L. Ableitung des Stefanschen Gesetzes, betreffend die Abhängigkeit der Wärmestrahlung von der Temperatur aus der elektromagnetischen Lichttheorie. Annalen der Physik, 22, 2 (1884), 291-294.

48   Rayleigh, B.J. W.S. The theory of sound. Macmillan and Company, London, 1896.

49   Planck, M. Über eine Verbesserung der Wien'schen Strahlungsgleichung. Verhandlungen der Deutschen Physikalischen Gesellschaft (1900).

50   Sato, T. Spectral Emissivity of Silicon. Japanese Journal of Applied Physics, 6 (1967), 339-347.

51   Timans, P.J. Emissivity of silicon at elevated temperatures. Journal of Applied Physics, 74, 10 (1993), 6353-6364.

52   Sopori, B. et al. Calculation of Emissivity of Si Wafers. Journal of Electronic Materials, 28, 12 (1999), 1385-1389.

53   Ravindra, N. M. et al.. Emissivity Measurements and Modeling of Silicon-Related Materials: An Overview. International Journal of Thermophysics, 22, 5 (2001), 1593-1611.

54   Ravindra, N. M. Modeling and Simulation of Emissivity of Silicon-Related Materials and Structures. Journal of Electronic Materials, 32, 10 (2003), 1052-1058.

55   Lee, Bong Jae and Zhang, Zhuomin. Rad-Pro User Manual. Atlanta, 2005.

56   Tada, H. Novel Microelectromechanical Systems (MEMS) for the Study of Thin Film Properties and Measurement of Temperatures During Thermal Processing. Tufts University. 1999.

57   Lambert, M. O. et al. Temperature dependence of the reflectance of solid and liquid silicon. Journal of Applied Physics, 52, 8 (1981), 4975-4976.

58   Shvarev, K. M., Baum, B. A., and Geld, P. V. Optical Propertie of Liquid Silicon. High Temperature, 15 (1977), 548.

59   Reichel, D. et al. Radiation Thermometry-Sources of Uncertainty during Contactless Temperature Measurement. In Skorupa, W. and Schmidt, H., eds., Subsecond Annealing of Advanced Materials. Springer, Heidelberg, 2013.

60   Qu, Y. et al. Insertion Error in LPRT temperature measurements. (Kyoto, Japan 2006), International Conference of Advance Thermal Processing of Semiconductors.

61   Larson, B. C. et al. Time-resolved x-ray Diffraction measurement of the temperature and temperature gradients in silicon during pulsed laser annealing. Applied Physics Letters, 42, 3 (1983), 282-284.

62   Furukawa, H. et al. In-situ measurement of temperature variation in Si wafer during millisecond rapid thermal annealing induced by thermal plasma jet irradiation. Japanese Journal of Applied Physics, 47, 4S (2008), 2460-2463.

63   Murakami, K. et al. Measurement of lattice temperature during pulsed-laser annealing by time-dependent optical reflectivity. Japanese Journal of Applied Physics, 20, 12 (1981), L867-L871.

64   Geiler, H. D. et al. Explosive crystallization in silicon. Journal of Applied Physics, 59, 9 (1985), 3091-3099.

65   Sedgwick, T. O. Pyrometric measurement of temperature during cw argon-ion laser annealingand the solid state regrowth rate of amorphous Si. Applied Physics Letters, 39, 3 (1981), 254-255.

66   Sampson, R. K. et al. Simultaneous silicon wafer temperature and oxide film thickness measurement in rapid-thermal processing using ellipsometry. Journal of the Electrochemical Society, 140, 6 (1993), 1734-1743.

67   Yamada, Y. and Ishii, J. In situ Silicon-Wafer Surface-Temperature Measurements Utilizing Polarized Light. International Journal of Thermophysics, 32, 11-12 (2011), 2304-2316.

68   Luxtron Corporation. LUXTRON Announces TrueTemp 7150 Ripple Technology, for Non-contact In situ, Real-time RTP Wafer Temperature Measurement in Lamp Heated Environments; Emissivity Independent Accurate Temperature Measurement Featuring LUXTRON's Patented Ripple Technology. Santa Clara, 1997.

69   Moslehi, M. M. Method and apparatus for precise temperature measurement. USA Patent US 5180226 A (1993)

70   Oh, M. et al. Impact of emissivity-independent temperature control in rapid thermal processing. (Boston 1997), Material Research Society.

71   Neubig, B. and Briese, W. Das Große Quarzkochbuch. Franzis Verlag, Feldkirchen, 1997.

72   Kleiner, M. Acoustics and Audio Technology. J. Ross Publishing, Fort Lauderdale, 2012.

73   Tipler, Paul A. and Mosca, Gene P. Physics for Scientists and Engineers. Palgrave Macmillan, 2007.

74   Bajaj, N.K. The Physics of Waves and Oscillations. Tata McGraw-Hill Publishing Company Limited, Delhi, 1984.

75   Smith, M. et al. Modeling and regrowth mechanisms of flash lamp processing of SiC-on-silicon heterostructures. Journal of Applied Physics, 96, 9 (2004), 4843-4851.

76   Chu, P.K.H. Quantitative analysis of semiconductor materials using secondary ion mass spectrometry. Cornell University, 1982.

77   Yan, Q., et al. Insertion Error in LPRT temperature measurements. (Kyoto 2006), International Conference of Advance Thermal Processing of Semiconductors.

78   Murray, J. L. and McAlister, A. J. The Al-Si (Aluminium-Silicon) System. Bulletin of Alloy Phase Diagrams, 5, 1 (1984), 74-84.

79   Rakić, A. D. Algorithm for the determination of intrinsic optical constants of metal films: application to aluminum. Applied Optics, 34, 22 (1995), 4755-4767.

80   Okamoto, H. and Massalski, T. B. The Au-Si (Gold-Silicon) System. Bulletin of Alloy Phase Diagrams, 4, 2 (1983), 190-198.

81   Palik, E. D. Handbook of Optical Constants of Solids. Academic Press, Boston, 1985.

82   Hunter, A. et al. Traceable emissivity measurements in RTP using room temperature reflectometry. In 11th International Conference on Advanced Thermal Processing of Semiconductors (Charleston, South Carolina, USA 2003).

83   Hackenberg, M. Personal Communications. May 2013

84   Bolze, D. Personal Communications. February 2013

Printed in the United States
By Bookmasters